# "INDUSTRIALIZATION"

## CHALLENGES & PANACEA FOR AFRICA; A CASE STUDY OF NIGERIA

Wake-up call to a slumbering giant to arise and pursue an aggressive & sustainable industrialization policy through the application of science and technology.

JOB LUKA IZANG. M.Sc

# DEDICATION

This book is dedicated to Jesus Christ and my two beautiful Queens Glorious - Keller And Trisha – Pearl.

# TABLE OF CONTENTS

PREFACE

CHAPTER 1:

EVIDENCE OF NIGERIA'S SLUMBER............................................8

1.1 Decay and Obsolescence..................................................22

1.2 Nigeria's De-industrialization..........................................23

CHAPTER 2:

EFFECTS AND DANGERS OF NIGERIA'S CONTINUOUS SLUMBERING........................................................................25

2.1. Decades of scientific and technological backwardness..........25

2.2. Energy transitions.......................................................62

2.3. The Negative Effects of Overpopulation............................71

2.4 Overtaken by previous equals..........................................82

CHAPTER 3:

ARRESTING THE VISCIOUS CIRCLE OF NIGERIA'S SLUMBERING........................................................................97

3.1. A Call for Nation Building.............................................97

3.2 Aggressive pursuit of STIR&D........................................111

3.3 Industrialization..........................................................126

# CHAPTER 4:

PANACEA TO OVERCOME NIGERIA'S PERSISTENT SLUMBERING..........145

4.1. National Economic Development Blueprint..........145

4.1.1. Overview of NIPF-Nv20:2020 and NSTIR 2017-2030...153

4.1.2 Proposed NRMP 2019-2040- Blueprint..........169

4.1.3 Re-design of Nigeria's Education system architecture..........188

4.1.3.1. Proposed Recommendations on education..........196

4.1.4. Proposed creation of ICSF..........200

4.1.4.1. Beyond Politics – Overcoming the barriers..........205

# CHAPTER 5:

RECCOMMENDATIONS FOR NIGERIA'S PERMANENT ALERT POSTURE..........208

5.1. Enforcing Multi-dimensional economic strategies..........208

5.2. Encouraging Home/garage Laboratories..........210

REFERENCES..........231

# PREFACE

**Industrialization** as defined by investopia is the process by which an economy is transformed from primarily agricultural to one based on the manufacturing of goods. Individual manual labor is often replaced by mechanized mass production, and craftsmen are replaced by assembly lines. Characteristics of industrialization include economic growth, more efficient division of labor, and the use of technological innovation to solve problems as opposed to dependency on conditions outside human control.

Africa's problem is multi-faceted and the only panacea to resolving them is taking the pathway to the realization of meaningful industrialization through the application of science and technology. Indigenous scientific Research and Development (R&D) is essential and indispensable to the industrialization of the continent.

**Awaken a sleeping giant** means to prompt a powerful force of unknown destructive capabilities into action. We know that the giant's heart is here, but his head is in dreamland.
Hitler awakened a sleeping giant by declaring war on the USSR.
Japan awakened a sleeping giant by bombing Pearl Harbor.
The slumbering giant that was China has finally awoken.
It is often said that when Admiral Yamamoto ordered the attack on Pearl Harbor, he awakened a sleeping giant. Even though the Nazis had occupied much of Europe, bombed London, and advanced to the outskirts of Moscow, the United States was still

slumbering on the morning of December 7, 1941.

The Cold War which followed on the heels of World War II kept America on its toes for four more decades, but after the fall of the Soviet system, the sleeping giant resumed its slumber. Then September 11, 2001 came suddenly enough —an attack which cost more American lives than Pearl Harbor. It was a classic wake-up call. And for a while, the giant awoke.

In the same vein, Nigeria as the sleeping giant of Africa needs to wake up now rather than latter to present day realities to avoid the devastation that comes with complacency and slumbering. A new vibrancy has been restored to the national psyche. Hope abounds not only in Nigeria but also on the entire continent. It seems, the sleeping giant is awakening and is now ready to take its position among the League of Nations. The signals look good and after a long dark period, the future looks bright. The fog has been lifted and we as a nation can see the road ahead, the road we must travel. We now seem to know where we are headed, but the journey is neither going to be smooth nor short. This may be a long and arduous journey to national reconstruction and fulfillment.

# CHAPTER 1:

## EVIDENCE OF NIGERIA'S SLUMBER

For many years Nigeria has been referred to as the Sleeping Giant of Africa. This derogatory reference has gained some legitimacy over the past decade due to the progressive decay of the nation's infrastructural base and the crumbling of its once booming economy. With a population of close to 190 million people, Nigeria enjoys the singular honor of being the most densely populated setting of black people in the world. Its multi-cultural and multi-ethnic groups have generated comparison to the United States of America. Indeed, with nearly 250 different ethnic groups, Nigeria is one of the most multi ethnic nations in the world. It is a country that is so immensely endowed with oil and other natural resources, that it is literally the envy of much of the world. Nigeria has been exceptionally favored and blessed with abundant natural, mineral and human resources. A nation that is totally devoid of natural calamities or disaster cannot be said to be less favored but the only natural disaster is bad leadership. Nigeria's has myriads of political problems (corruption-civil or social), lawlessness and deceit in high places. Nigeria also boasts of one of the most educated and active intelligentsia of any nation. Unfortunately, despite its natural and human endowments, corrupt and inept leadership has plagued Nigeria for many years. Despite enormous earnings from vast holdings in oil and other natural resources, Nigeria has witnessed an unprecedented decline in quality of life and an infrastructural decay over the past 25 years as its economic lifeblood is literally sucked away by a corrupt and greedy upper-class.

Previously, economic development had been hindered by years of military rule, corruption, and mismanagement. During the oil boom of the 1970s, Nigeria accumulated a significant foreign debt to finance major infrastructural investments. With the fall of oil prices during the 1980s oil glut Nigeria struggled to keep up with its loan payments and eventually defaulted on its principal debt repayments, limiting repayment to the interest portion of the loans. Arrears and penalty interest accumulated on the unpaid principal which increased the size of the debt. However, after negotiations by the Nigeria authorities, in October 2005, Nigeria and its Paris Club creditors reached an agreement in which Nigeria repurchased its debt at a discount of approximately 60 per cent. It used part of its oil profits to pay the residual 40 per cent, freeing up at least $1.15 billion annually for poverty reduction programmes. Petroleum plays a large role in the Nigerian economy, accounting for 40 per cent of GDP and 80 per cent of government earnings. However, agitation for better resource control in the Niger Delta, its main oil producing region, has led to disruptions in oil production and prevents the country from exporting at 100 per cent capacity. Although it is a producer of oil, the country has no functioning refinery. Nigeria is one of the poorest oil producing countries, according to the United Nations report citing "the economic policy orientation during the 70s left the country ill prepared for the eventual collapse of oil prices in the first half of the 80s. "Public investment was concentrated in costly and often inappropriate infrastructure projects with questionable rates of return and sizeable recurrent cost implications while the agricultural sector was largely neglected Nigeria also has a wide array of underexploited mineral, which include natural gas, coal, bauxite, tantalite, gold, tin, iron ore, limestone, niobium, lead and zinc. Despite huge deposits of these natural resources, the mining industry in Nigeria is still in its infancy. About 60 per cent of Nigerians work in the agricultural sector, and Nigeria has vast areas of underutilized arable land. The collapse of the manufacturing sector has led to mass unemployment. Corruption is on the increase in Nigeria. In 2009, it was listed among the failed states. Infrastructure is at low ebb and ethnic and religious tensions have unleashed the fear of peaceful co-existence. Nigeria's successive governments since political independence in 1960 all pledged increased commitment to use science and technology to enhance Nigeria's development. Nigeria's persistent impoverishment shows that in practice none of them succeeded. Consequently, it has been painful to remain only spectators as the advanced nations continue to employ science and technology to address their economic and social needs. Nigeria is politically an unstable country. The country has been bedeviled by all sort of negative forces: There is inadequate business infrastructure, power supply is erratic, and there is limited supply of clean treated drinkable water. There is periodic fuel scarcity, inadequate crime control and weak judicial system. The quality of our education system is also a problem. Our education system has deteriorated to a point where students cannot acquire the necessary skills they need to become employable or innovative in an ever-changing world, upon graduation. Higher institutions are plagued with inadequate Science and Technology facilities and materials for practical skills development. Many laboratories lack the basic equipment for thorough scientific research. How, for example, can a computer science graduate not understand the basics of writing software codes? Nigeria is churning out thousands of S&T graduates each year, but several of them are under-employed, going into banking and other non-scientific sectors. There is no doubt that human capital is the most important form of wealth for a modern nation. Economically successful countries are those that are able to turn technical innovation into economic productivity. Effective Science and technology (S&T) policies are thus crucial for developing countries. Nigeria as a developing nation must appreciate the potency of S&T to bring about significant changes in our local, state and national lives. The structure of the Nigerian economy is typical of an underdeveloped country. Data available has shown that between 2011 and 2012, the primary sector, in particular the oil and gas sector, dominated GDP, accounting for over 95 per cent of export earnings and about 85 per cent of government revenue, while in 2011, the manufacturing sector contributed only 4 per cent to GDP.

Nigeria is the largest economy in Africa; in 2013 its population was in excess of 170 million, with GDP of over US$500 billion (World Bank 2014). The continent's biggest oil exporter is also home to large natural gas reserves. The economy has recorded considerable acceleration in growth; real GDP grew by 6.3 per cent, 7.6 per cent, and 7.4 per cent in 2009, 2010, and 2011 respectively. Despite this, poverty is persistently high, and the structure of the economy is that of a typically underdeveloped country.

Over the years, our consumption rate has tremendously increased yet our capacity to be self-reliant has dwindled. The Nomura Food Vulnerability Index ranks Nigeria as the 4th most vulnerable country to global food price shocks, out of 80 countries. We lack the right plant varieties and storage systems to be efficient. The current infrastructure base in Nigeria is grossly inadequate in terms of capacity and quality. Power generation capacity is less than 4,000 MW—about 20 per cent of estimated national demand. A key challenge for government and the private sector is to build a modern, efficient, and effective infrastructure network within the next five to ten years.

Nigeria Compared with other Nations from the CIA World Fact

book 2009, the followings facts were obtained. The life expectancy (LEB) of people in Nigeria by 2009 statistic was 46.94 years quote placing us in the 209th position with Japan having 82.12 years LEB. Nigeria was ranked 8th in population density with a population of 149,229,090 people. The percentage of people below poverty line was 70% placing us in the 9th worst country in the world while Taiwan has less than 1%. The industrial production rate of Nigeria was 2.8% placing us in the 87th position as against China (9.3%) in the 15th position. Electricity production by Nigeria was 22.11bKwh placing us in the 68th position with the United States (4,167 bKwh) in the 1st position. South Africa with one third of the population of Nigeria had 264 bKwh of electricity production. Oil production was 2,440,000 barrels per day placing us at the 12th oil producing country in the world although fuel was still imported to power the economy. Nigeria's gas production was put at 21,480,000,000 cubic meters placing us as the 26th producer in the world although gas was out of reach of the poor. The inflation rate of 11.6% in the period 2000-2010 placed the country in the 56th worst position. The growth rate in GDP of 5.3% placed the country in the 76th position. The GDPPC of $2,100 also placed us in the 175th place. Nigeria's current account balance was $14,610,000,000 and foreign reserves of $50,330,000,000 placed us in the 20th and 26th positions respectively. Nigeria's total export value in 2009 was $76.8b. This placed us at the 42nd position. According to the Daily

Trust of November 26, 2008 in Dike (2009) the federal government acknowledged that about 80 per cent of Nigeria's youths are unemployed and 10 per cent underemployed.

For Nigeria to uplift herself successfully from her current state of underdevelopment, she too must make a deliberate effort to incorporate science into her culture and apply technology to raise the standard of living of her people. The ability to engage fruitfully in science and technology is not the birthright of any particular country or people, and yet too many people have the erroneous mentality that science and technology belongs to the advanced nations who are ordained to ration it out to developing nations as they see fit. We abhor the endemic corruption that has infested Nigeria and the damage it has done to retard development in our society. There is no doubt that corruption will make it difficult not just for science and technology to thrive, but for any other developmental effort to have a chance of succeeding. Despite the size and fast pace of economic growth in the Nigerian economy over the last decade, the manufacturing sector remains weak. Past policy efforts aimed at improving the performance of the sector have failed, and the focus has shifted towards more targeted policies aimed at specific sectors, A number of challenges exist that will be critical to the success or failure of this strategy. Key among these is infrastructure, corruption, and national security Nigeria ranks highly in the Corruption Perception Index. This has implications for investment and foreign direct investment (FDI) flows into the country. Previous anti-corruption policies implemented in Nigeria have been targeted at enforcement measures rather than addressing the root causes, which include social insecurity and over-centralization of resources. Successive Nigerian administrations since political independence have not given our schools and science education the type of support they need in order to be effective. As a result, we do not have enough indigenous educated manpower such as scientists, engineers, technologists et cetera that can grow our science and technology and make it as competitive as anyone's, in spite of the fact that Nigeria is blessed with an abundant human resources and wealth from natural resources. The current state of Nigeria is a clear indication that science and technology cannot thrive in a country where most of the schools are sub-standard and have no laboratories or adequately trained science teachers. The internal security of Nigeria has become a major challenge. Internal conflicts, including religious, ethnic, and economic, have had debilitating effects on the economy. The insecurity of lives and properties became noticeable following the civil war and the subsequent military regimes, and the recent upsurge of violence and insurgency in the country heightens the need to address the persistent causes of social tension

Nigeria as a country is undoubtedly struggling with a long list of contemporary issues. The government currently has a lot on its shoulders, so the responsibility of solving social problems partly lies on the citizens. A complete analysis and description of all the primary contemporary social problems in Nigeria that are influencing the country's population and waiting for wise solutions are listed below.

**National identity problem**; Nigeria is a relatively young country because it gained independence only 57 years ago, in 1960. It was also created in an artificial way after gaining freedom from its colonial masters. There is a large number of tribes in Nigeria, and this is one of the important issues being faced, because to this day the country still has problems with gaining and voicing a national idea. The main purpose of this idea of nationality is to unite all the social, religious and ethnic groups within one country and make them identify as Nigerians despite all their differences. Besides, there are some conflicts with other nations that create unnecessary tension among Nigerians.

**Poverty of the population;** This is, without any doubt, a large social issue which does not portray Nigeria in the best light. Poverty causes a lot of problems in the society, and while the population of the country is huge, a large part of it does not have access to the basic needs for comfortable living. Statistical data shows that about 70% of the country's residents live below the line of poverty, and there are a lot of people who are forced to survive on just $2 per day, which is a critical situation. The population is rapidly growing, and soon, even more people will be under the risk of living in poor conditions without money and food.

**Corruption among influential branch of Nigeria;** Nigeria is known as a country with a high level of corruption amongst its government representatives. Reports show that some people in government offices earn as much in a year, as the average Nigerian could earn in 65 years; this shows that they definitely have other sources of income apart from their official salary. In all honesty, the governmental system in the country is corrupt and there is a clear lack of justice. This has led to an increased level of stealing, manipulation and bribes. Corruption is the main reason for class division; while people living in poverty are struggling to survive, the rich people have huge possessions and are becoming wealthier with every passing year.

**Inequality between wealthy and poor;** This problem was briefly pointed out in the last paragraph – in fact, Nigeria is widely called "the rich country with poor people". The reason for this is a huge amount of natural resources all over the country which are not equally spread amongst citizens – the bigger quantities go to the rich people, while the poor ones are left to suffer. In light of the poverty problem, it is quite ironic to know that some of the richest people in Africa come from Nigeria. The primary issue is that the main part of the country's population is involved in petty agriculture and earn very little. Only a few chosen people have access to the oil production sector, which brings the biggest income to the country. This leaves the situation as it is – poor people are destined to struggle with poverty, while the rich cannot get enough of their expensive belongings.

**Terrorist attacks;** The terrorist group called Boko Haram is one of the reasons for Nigeria's worldwide fame. This organization is fighting against the western way of life and they have already murdered hundreds of citizens during their terrorist attacks. Lots of people have lost their homes or gotten displaced from their hometowns due to the activities of Boko Haram. This issue is definitely an important one – terrorists have to be dealt with before they can kill more innocent residents.

**Child mortality;** Contemporary issues in Nigeria are all serious, however, this one really hits home. According to the statistics, over 2000 children die of hunger, diseases and abandonment daily in Nigeria. Nigeria comes second in the world's child mortality rating. The leading reasons for this huge issue are lack of education amongst women, bad health care system in the country and, most importantly, general poverty of the population. As already mentioned, a huge part of Nigerians is forced to live on $2 per day or even less, so it was estimated that $10 could save the life of one child because this is the price of antibiotic medicine. Unfortunately, poor parents cannot afford such a luxury, and a lot of children under the age of 5 die every day in Nigeria.

**High rate of unemployment;** Most of the unemployed in the country are those who left their homes to search for a better life;

such movements are the reason major cities like Lagos is overpopulated – people go there to look for opportunities. This is a strong social issue which often leads to psychological problems. The current rate of unemployment in Nigeria is about 8.2%.

**Low level of education;** Lots of people have strong negative opinions about Nigeria's system of education, and they mostly blame the government and their involvement with corruption for this. The statistics show that only 50% of women and 70% of men in Nigeria can read or write. This is obviously an issue of high importance because our world is based on information technologies and uneducated people do not stand a chance of succeeding in this world.

**Division of tribes;** Tribal conflicts are common in Nigeria because the country is young and has not existed for too long. The biggest tribes in Nigeria are Hausa, Igbo and Yoruba, and they all are culturally and religiously different, not to mention the fact that they all speak their own language. These differences have been causing conflicts all through the years, and this situation is yet to change. The different tribes still have conflicts with each other from time to time, and this does not help in making sure they remain one Nigeria. There are a lot of contemporary social issues in Nigeria, however, the country is not completely hopeless – the mindset of the government and people can still be changed. It may take a long time, but there is still a chance that these problems can be dealt with if we work hard enough.

Researchers should be celebrated for their superior intellectual abilities with respect to their endowed national development potentials; but reverse is the case in our society today, rather, more focus are given to the so-called entertainers mostly the half-exposed ones. How will intellectual revolution be encouraged in Nigeria since scientists, inventors, innovators, technologists and researchers are given less recognition?

In fact, I'm not against entertainment per say, but more importance shouldn't be given to the sector especially for a country like Nigeria that is seriously yearning for development. We need to get our priorities right in this country!!! Respect should be given to superior intellectual resources in order to encourage a scientific and technological development in Nigeria, since a country's development index is being rated by her level of science and technology!

Let's ask ourselves; In Nigeria, why do we import the products rather than importing the machines. We have been importing toothpicks, why can't we import the toothpick-making machine and get the jobless youths employed? And this will also conserve the much needed foreign exchange for importation of these cheap products.

So N-Power now has 1 million applicants within few days that are good gesture by our FG! That's social security and empowerment! That also shows the level of unemployment in this country! But will this monthly stipend last for forever? Let us call a spade a spade and not a farm-tool; this will not solve this jobless situation of the country! It's a nice move by the FG and also structured as a poverty alleviation scheme. But I think it's better to ignite an industrial revolution in Nigeria and let the youths be actively involved in the nation building process. That will solve our recurring problems of unemployment and wailing poverty and will also increase the standard of living of the citizenry. FG should therefore establish manufacturing industries across the 6-geo political zones of Nigeria. Most aircrafts that operate in Nigeria perform their routine maintenance services abroad, what stops us from having our own indigenous facilities being established in Nigeria in order to maintain those aircrafts?? We have jobless engineers in this country, how will they practice what they studied? How will they innovate and invent solutions?? Before Japan experienced her industrial revolution, several of her citizens were sent all over advanced countries in order to have modern education, technical skills and also advanced technological exposures. Those citizens got back to their country to ignite an industrial revolution,

so why can't Nigeria take a leaf from Japan, China and the Asian Tigers?

We also need to recognize the fact that the first step of a nation building process is to unite the diverse nationalities into a beautiful rainbow nation with a unique national identity, in order to peacefully coexist for the pursuit of a sustainable national development agenda. "Build these people, and these developed People will successfully build a nation". The Chinese people recognized the strength in their diversity and also discovered the advantage of their large population; those citizens are being massively used as workforce in the Chinese industries for cheap labor. When will Nigeria maximize the large population in the country to blaze a trail for an industrial revolution?

## 1.1 Decay and Obsolescence:

Nigeria's past developmental programmes which were initiated, designed and constructed during the early sixties and late seventies by the country's founding fathers, are now a shadow of its old self glory. These initiatives are currently wallowing in a prolonged fallow period marred by consistent past maladministration. Majority of those past industries which are created by the good intentions of Nigeria's founding fathers for the purpose of nation building have become failures, several decades latter. Being the symbol of her strength and wealth as a young, vibrant rising nation have been neglected over the years. This is a very strong exhibition of the effects of continuous slumbering by Nigeria as Africa's potential giant which has everything in its capacity and at its disposal to be a model country worthy of emulation by the rest of Africa and other nations of the world.

Some of the industries have long been forgotten through extinction like the defunct Nigerian Telecommunications Limited(NITEL), Nigerian Airways (NA), Nigeria Electricity Power Authority ( NEPA), while others have deteriorated and become dysfunctional with obsolete and outdated equipments to the point of scrap value like the Ajaokuta Steel Rolling Mill, The Zuma Steel Rolling Mill( former Jos Steel Rolling Mill), Jos International Breweries Limited ( JIB), Limca Bottling Company Limited, while others yet are shadows of their old selves and are crawling along like proverbial lambs ready for the gallows or slaughter like the three Nigerian refineries at Wari, PortHarcourt and Kaduna originally intended to serve both internal and external consumptions, the Fertilizer Manufacturing Company of Nigeria ( FERMACO), just to mention but a few. The Ajaokuta steel complex has been gestating for more than 30 years despite the billions of Naira invested, while similar projects have been completed in three to five years in some countries.

## 1.2 Nigeria's De-industrialization

On the other side of the spectrum, the textile industry is an example of a sunset industry and illustrates the deindustrialization process that Nigeria has experienced in the last two decades. Over 820 companies shut down or suspended production between 2000 and 2008 (MAN 2009). At its peak, the textile industry employed close to 700,000 people (making it the second largest employer after the government) and generated a turnover of over US$8.95 billion. The industry witnessed a catastrophic collapse, from 175 firms in the mid-1980s to ten factories in stable condition in 2004, while employment in the industry plunged from 350,000 to 40,000.

## CHAPTER 2:

## EFFECTS AND DANGERS OF NIGERIA'S CONTINUOUS SLUMBERING

### 2.1 Decades of scientific and technological backwardness

Amongst Nigeria's numerous and persistent problems, one of the most pressing is her backwardness in science and technology. According to Iredale (2003), countries with the most intellectual resources achieve the highest rates of economic growth and the fastest development in Science and Technology (S&T). Knowledge produces economic riches and also is a vital ingredient for dealing with many of the social and environmental problems of our lives today. Industrialized countries (United States, the United Kingdom, Canada, Germany, Japan, Singapore, Hong Kong and Australia), give priority to policies aimed at attracting highly skilled immigrants.

Science and technology is the expression of man's highest ingenuity and man's ingenuity can only be expressed optimally in a suitable environment which shamefully is lacking in Nigeria today.

While the Nigerian government is engrossed in strategies to

revive the country's moribund steel industries, which have been in a stagnated state of production, most industrialized nations are already fine-tuning advanced technologies for the production of newer materials such as composites to replace steel and aluminum. Composites are highly advanced materials; some constructs are stronger than steel pound for pound and, yet, are lightweight. They also have high corrosion resistance, low conductivity, and many other attributes that make them indispensible in most industrial applications. Lighter and stronger materials can directly translate into more efficient systems and engineering constructs such as faster but more eco-friendly airplanes, ships, trains, and cars. Its extensive use in civil construction, energy, military, and other vital sectors of national strategic importance directly impact the economic fortunes and technological competitiveness of countries on the global stage, signaling the advent of a new revolutionary era. Effective science and technology policies are crucial in advancing Nigeria economically, socially and politically given the varieties of challenges facing the largest populated nation in Africa. Alternate sources of energy to alleviate the inadequacy of electricity supply in Nigeria are suggested base on availability of needed local resources. Globalization is driven by science and technology. IT is the engine of economic liberalization and associated developments in international trade. According to Ogbu (2004), Nigerians as consumers of science and technology are fascinated by gadgets especially the latest ones, their speeds and designs/brands are appreciated. Science and technology are central to the developmental prospects of poor countries. They can provide tools that help alleviate the specific problems that afflict many poor countries and which impede their development prospects, such as disease, infrastructural (energy, communication, transport, etc.) decay, and the degradation of the environment. S&T are also central to the dynamics of economic development itself. Economically successful countries are those that are able to turn technical innovation into economic productivity. The success stories of Japan, Korea, and Taiwan, for example are in large part stories of a long-term strategic policy focused on fostering indigenous innovation capacity.

Recent technological developments including the printing press, the telephone, the computer and the Internet have lessened physical barriers to communication and allowed humans to interact freely on a global scale. Esho (2008) emphasizes that the strength of a nation is gauged and evaluated on the basis of contemporary issues and challenges left unresolved. The achievements of a nation is a measure of the commitment to technical education in terms of her technological contribution to global economy and how national needs are matched with technical education curricula and proper planning.

Compton (2004) mentioned that due to the nature of contemporary society, the relationship between the domains of science and technology has never been stronger. Scientific knowledge and methodologies themselves provide a major source of input into the development of technological practices and outcomes. They are also key tools in the establishment of explanations of why technological interventions were, or were not, successful. Technological practices, knowledge and outcomes can provide mechanisms for science to gain a better view of its defined world, and in fact can provide serious challenges to the defining of that world. For example, the development of the technological artifacts that extend the observation capabilities of humans (such as the telescope and microscope), made 'visible' and available 'new worlds' for science to interrogate and explain.

Ogbu (2004) further asserts that Africa's brain drain phenomenon has both pull and push factors that have contributed significantly to the poor state of Science and technology in the region. Many top scientists left the shores of Nigeria and refused to return because some developed countries also put in place policies to attract highly specialized Africans thereby depleting the meager stock.

Every year the Noble peace prize award was given out to outstanding personalities who excel in their field of endeavor. Some in the field of science, technology, physics, Biology, chemistry, space technology and exploration, Medicine, pharmaceuticals science, engineering etc. through innovative inventions, theoretical postulations, literary work and in peace advancement around the world. This is the highest award that could be granted to any individual or group for their contribution to global development.

Ideas, it is said rule the world. Scientific and technological advancement brought about through ideas and conceptual intellect has raised nations who were hitherto once seen as hopeless but are now major global movers. For instance, China was invaded by Japan during Second World War and looked devastated and hopeless mostly in ruins after the war, but immediately embraced technological and scientific approach and today the story is different. Japan has no major natural resources except lumber, but through "Good thinking, good product" policy initiative has adopted science and technology as its major economic drive. Today, Japan is the $3^{rd}$ largest economy in the world.

Nigeria must arise from her slumber and also embrace science and technology (just as other countries did) as the bedrock of her economic and developmental growth, if it must be counted amongst the committee of nations that build their foundation on a sound propelling force.

One of the most important reasons why Nigeria must arise is that Nigeria is left behind by approximately seventy decades to more than a century of scientific and technological backwardness, what this means is that before Nigeria could catch up in scientific and Technological advancement (STA) and Science, Technology, Engineering and Mathematics (STEM) to be at par with countries like USA, Germany, France,

UK, Russia, China, Japan etc it will take Nigeria a minimum of one Century or more because these are the most advanced and industrialized countries in the world. The second category is the so called "third world countries" who are next in affluence to the most industrialized countries some of these are India, Brazil, Hong Kong, south Korea, Malaysia, Singapore, Indonesia etc. For this category, it will take Nigeria approximately forty to seventy decades of aggressive and consistent STA to be on a level footing or at par with them. This is so, because the same countries themselves takes this much time, more or less to reach where they are today, with persistent determination, stability in government, consistency in industrialization drive and blue prints and pluralistic government reform agenda. Whatever arrears period these countries have left Nigeria behind, they are well established and near efficient in power generation of various kinds including renewable and non – renewable energy sources, advanced and updated industrial plants, machines and equipments, state of the art research laboratories and development centers, well developed manufacturing sector, highly skilled manpower, favorable business and investment climate.

Closing this gap for Nigeria will take sincere commitment on the part of the country's leadership to do so. This will take will power, determination, relevant legislation, funding, enforcement, endurance and consistency. The challenge at hand for Nigeria to achieve such scientific and technological advancement is for Nigeria to organize its scientists, technologists, innovators, inventors and industrialists both far and near to come back home from diasporas and contribute their quota into nation building of their motherland through funding investments, input of new ideas, injection of advance foreign acquired skills. Good management and leadership qualities will encourage all to imbibe patriotism for their mother land and to instill confidence on them to believe that they can do all things if they believe and wish to.

Nigeria is left behind because it has become a dumping ground for obsolete equipments. For example, Laptops and desktops computers with 4GB RAM is still considered as massive and comfortable in Nigeria while in the advance world, they are working with 2 –3 Terabytes of RAM. Nigeria has no AI in any of its programme, while the advanced countries has gone as far as making robots, space crafts, driverless vehicles, orbiting satellites, cruise missiles and super-computers to have and operate on Artificial intelligence which enables the machines to have the capacity for independent thinking, judgment and decision making on their own as well as logical engagement.

Past, present & future Inventions and innovations that shaped the nature of modern life;

(a) Past Science and Technology Chronological Inventions:

I do not think we can understand the contemporary world without understanding the events that have given rise to it. Inventions don't generally happen by accident or in a random order: science and technology progress in a very logical way, with each new discovery leading on from the last. It can be seen from the mini chronology of invention, below.

**Prehistory**

*10 million years ago;* Humans make the first tools from stone, wood, antlers, and bones.

*1–2 million years ago;* Humans discover fire.

*25,000– 50,000 BCE;* Humans first wear clothes.

*10,000 BCE;* Earliest boats are constructed.

*8000– 9000 BCE;* Beginnings of human settlements and agriculture.

*6000– 7000 BCE;* Hand-made bricks first used for construction in the Middle East.

*4000 BCE;* Iron used for the first time in decorative ornaments.

*3500– 5000 BCE;* Glass is made by people for the first time.

*3500 BCE;* Humans invent the wheel.

*3000 BCE;* First written languages are developed by the Sumerian people of southern Mesopotamia (part of modern Iraq).

*2500 BCE;* Ancient Egyptians produce papyrus, a crude early version of paper.

*3000– 600BCE;* Bronze Age: Widespread use of copper and its important alloy bronze.

*2000 BCE;* Water-raising and irrigation devices like the shaduf (shadoof), invented by the Ancient Egyptians, introduce the idea of lifting things using counterweights.

*C1700 BCE;* Semites of the Mediterranean develop the alphabet.

*1000 BCE;* -Iron Age begins: iron is widely used for making tools and weapons in many parts of the world.

*250 BCE;* Ancient Egyptians invent lighthouses, including the huge Lighthouse of Alexandria.

*600 BCE;* Thales of Miletus discovers static electricity.

*C.150– 100 BCE;* Gear-driven, precision clockwork machines (such as the Antikythera mechanism) are in existence.

*50 BCE;* Roman engineer Vitruvius perfects the modern, vertical water wheel.

*62 CE;* Hero of Alexandria, a Greek scientist, pioneers steam power.

*105 CE;* Ts'ai Lun makes the first paper in China.

*27 BCE–395 CE*; Romans develop the first, basic concrete called pozzolana.

**Middle Ages.**

*600 CE;* Windmills are invented in the Middle East.

*700–900 CE;* Chinese invent gunpowder and fireworks.

*800–1300 CE;* Thanks to inventors such as the Banū Mūsā brothers and al-Jazari, the Islamic "Golden Age" sees the development of a wide range of technologies, including ingenious clocks and feedback mechanisms that are the ancestors of modern automated factory machines.

**1000 CE?** Chinese develop eyeglasses by fixing lenses to frames that fit onto people's faces.

*1206;* Arabic engineer al-Jazari invents a flushing hand-washing machine, one of the ancestors of the modern toilet.

**1450;** Johannes Gutenberg pioneers the modern printing press, using rearrange able metal letters called movable type.

*1470s*; the first parachute is sketched on paper by an unknown inventor.

## 16th century

*1530s;* Gerardus Mercator helps to revolutionize navigation with better mapmaking.

*1590;* A Dutch spectacle makers named Zacharias Janssen makes the first compound microscope.

*1596;* Sir John Harington describes one of the first modern flush toilets.

## 17th century

*1600;* Galileo Galilei designs a basic thermometer.

*1600;* William Gilbert publishes his great book De Magnete describing how Earth behaves like a giant magnet. It's the beginning of the scientific study of magnetism.

*1609;* Galileo Galilei builds a practical telescope and makes new astronomical discoveries.

*Mid-17th century;* Antoni van Leeuwenhoek and Robert Hooke independently develop microscopes.

*1643;* Galileo's pupil Evangelista Torricelli builds the first mercury barometer for measuring air pressure.

*1650s;* Christiaan Huygens develops the pendulum clock (using Galileo's earlier discovery that a swinging pendulum can be used to keep time).

*1687;* Isaac Newton formulates his three laws of motion.Motion.

## 18th century

*1700s;* Bartolomeo Cristofori invents the piano.Pianos

*1701;* English farmer Jethro Tull begins the mechanization of agriculture by inventing the horse-drawn seed drill.

*1703;* Gottfried Leibniz pioneers the binary number system now used in virtually all computers.

*1712;* Thomas Newcomen builds the first practical (but stationary) steam engine.

*1700s;* Christiaan Huygens conceives the internal combustion engine, but never actually builds one.

*1737;* William Champion develops a commercially viable process for extracting zinc on a large scale.

*1757;* John Campbell invents the sextant, an improved navigational device that enables sailors to measure latitude.

*1730s– 1770s;* John Harrison develops reliable chronometers (seafaring clocks) that allow sailors to measure longitude accurately for the first time.

*1751;* Axel Cronstedt isolates nickel.

*1756;* Axel Cronstedt notices steam when he boils a rock—and discovers zeolites.

*1769;* Wolfgang von Kempelen develops a mechanical speaking machine: the world's first speech synthesizer.

*1770s;* Abraham Darby III builds a pioneering iron bridge at a place now called Ironbridge in England.

*1780;* Josiah Wedgwood (or Thomas Massey) invents the pyrometer.

*1783;* French Brothers Joseph-Michel Montgolfier and Jacques-Étienne Montgolfier make the first practical hot-air balloon.

*1791;* Reverend William Gregor, a British clergyman and amateur geologist, discovers a mysterious mineral that he calls menachite. Four years later, Martin Klaproth gives it its modern name, titanium.

## 19th century

*1800;* Italian Alessandro Volta makes the first battery (known as a Voltaic pile).

*1801;* Joseph-Marie Jacquard invents the automated cloth-weaving loom. The punched cards it uses to store patterns help to inspire programmable computers.

*1803;* Henry and Sealy Fourdrinier develop the papermaking machine.

*1806;* Humphry Davy develops electrolysis into an important chemical technique and uses it to identify a number of new elements.

*1807;* Humphry Davy develops the electric arc lamp.

*1814;* George Stephenson builds the first practical steam

locomotive.

*1816;* Robert Stirling invents the efficient Stirling engine.

*1820s– 1830s;* Michael Faraday builds primitive electric generators and motors.

*1827;* Joseph Niepce makes the first modern photograph.

*1830s;* William Sturgeon develops the first practical electric motor.

*1830s;* Louis Daguerre invents a practical method of taking pin-sharp photographs called Daguerreotypes.

*1830s;* William Henry Fox Talbot develops a way of making and printing photographs using reverse images called negatives.

*1830s– 1840s;* Charles Wheatstone and William Cooke, in England, and Samuel Morse, in the United States, develop the electric telegraph (a forerunner of the telephone).

*1836;* Englishman Francis Petit-Smith and Swedish-American John Ericsson independently develop propellers with blades for ships.

*1839;* Charles Goodyear finally perfects a durable form of rubber (vulcanized rubber) after many years of unsuccessful experimenting.

*1840s;* Scottish physicist James Prescott Joule outlines the theory of the conservation of energy.

*1840s;* Scotsman Alexander Bain invents a primitive fax machine based on chemical technology.

*1849;* James Francis invents a water turbine now used in many of the world's hydropower plants.

*1850s;* Henry Bessemer pioneers a new method of making steel in large quantities.

*1850s;* Louis Pasteur develops pasteurization: a way of preserving food by heating it to kill off bacteria.

*1850s;* Italian Giovanni Caselli develops a mechanical fax machine called the pantelegraph.

*1860s;* Frenchman Étienne Lenoir and German Nikolaus Otto pioneer the internal combustion engine.

*1860s;* James Clerk Maxwell figures out that radio waves must exist and sets out basic laws of electromagnetism.

*1860s;* Fire extinguishers are invented.

*1861;* Elisha Graves Otis invents the elevator with built-in safety brake.

**1867;** Joseph Monier invents reinforced concrete.

*1868;* Christopher Latham Sholes invents the modern typewriter and QWERTY keyboard.

*1876;* Alexander Graham Bell patents the telephone, though the true ownership of the invention remains controversial even today.

*1870s;* Thomas Edison develops the phonograph, the first practical method of recording and playing back sound on metal foil.

*1870s;* Lester Pelton invents a useful new kind of water turbine known as a Pelton wheel.

*1877;* Thomas Edison invents his sound-recording machine or phonograph—a forerunner of the record player and CD player.

*1877;* Edward Very invents the flare gun (Very pistol) for sending distress flares at sea.

*1880;* Thomas Edison patents the modern incandescent electric lamp.

*1880;* Pierre and Paul-Jacques Curie discover the piezoelectric effect.

*1880s;* Thomas Edison opens the world's first power plants.

*1880s;* Charles Chamberland invents the autoclave (steam sterilizing machine).

*1880s;* Charles and Julia Hall and Paul Heroult independently develop an affordable way of making aluminum.

*1880s;* Carrie Everson invents new ways of mining silver, gold, and copper.

*1881;* Jacques d'Arsonval suggests heat energy could be extracted from the oceans.

*1883;* George Eastman invents plastic photographic film.

*1884;* Charles Parsons develops the steam turbine.

*1885*; Karl Benz builds a gasoline-engined car.

*1886;* Josephine Cochran invents the dishwasher.

*1888;* Friedrich Reinitzer discovers liquid crystals.

*1888*; John Boyd Dunlop patents air-filled (pneumatic) tires.

*1888;* Nikola Tesla patents the alternating current (AC) electric induction motor and, in opposition to Thomas Edison, becomes a staunch advocate of AC power.

*1899;* Everett F. Morse invents the optical pyrometer for measuring temperatures at a safe distance.

*1890s;* French brothers Joseph and Louis Lumiere invent movie projectors and open the first movie theater.

**1890s;** German engineer Rudolf Diesel develops his diesel engine—a more efficient internal combustion engine without a sparking plug.

*1894;* Physicist Sir Oliver Lodge sends the first ever message by radio wave in Oxford, England.

*1895;* German physicist Wilhelm Röntgen discovers X rays.

*1895;* American Ogden Bolton, Jr. invents the electric bicycle.

## 20th century

*1901;* Guglielmo Marconi sends radio-wave signals across the Atlantic Ocean from England to Canada.

*1901;* The first electric vacuum cleaner is developed.

*1903;* Brothers Wilbur and Orville Wright build the first engine-powered airplane.

*1905;* Albert Einstein explains the photoelectric effect.

*1905;* Samuel J. Bens invents the chainsaw.

*1906;* Willis Carrier pioneers the air conditioner.

*1906;* Mikhail Tswett discovers chromatography.

*1907;* Leo Baekeland develops Bakelite, the first popular synthetic plastic.

**1907;** Alva Fisher invents the electric clothes washer.

*1906-8;* Frederick Gardner Cottrell develops the electrostatic smoke precipitator (smokestack pollution scrubber).

*1908;* American industrialist and engineer Henry Ford launches the Ford Model T, the world's first truly affordable car.

*1909;* German chemists Fritz Haber and Zygmunt Klemensiewicz develop the glass electrode, enabling very precise measurements of acidity.

*1912;* American chemist Gilbert Lewis describes the basic chemistry that leads to practical, lithium-ion rechargeable batteries (though they don't appear in a practical, commercial form until the 1990s).

*1912*; Hans Geiger develops the Geiger counter, a detector for radioactivity.

*1919;* Francis Aston pioneers the mass spectrometer and uses it to discover many isotopes.

*1920s*; John Logie Baird develops mechanical television.

*1920s;* Philo T. Farnsworth invents modern electronic television.

*1920s;* Robert H. Goddard develops the principle of the modern, liquid-fueled space rocket.

*1920s;* German engineer Gustav Tauschek and American Paul Handel independently develop primitive optical character recognition (OCR) scanning systems.

*1920s;* Albert W. Hull invents the magnetron, a device that can generate microwaves from electricity.

*1921;* Karel Capek and his brother coin the word "robot" in a play about artificial humans.

*1921;* John Larson develops the polygraph ("lie detector") machine.

*1928;* Thomas Midgley, Jr. invents coolant chemicals for air conditioners and refrigerators.

*1928;* the electric refrigerator is invented.

*1930s;* Peter Goldmark pioneers color television.

*1930s;* Laszlo and Georg Biro pioneer the modern ballpoint pen.                *1930s;* Maria Telkes creates the first solar-powered house.

*1930s;* Wallace Carothers develops neoprene (synthetic rubber used in wetsuits) and nylon, the first popular synthetic clothing material.

*1930s;* Robert Watson Watt oversees the development of radar.

*1930s;* Arnold Beckman develops the electronic pH meter.

*1931;* Harold E. Edgerton invents the xenon flash lamp for high-speed photography.

*1932;* Arne Olander discovers the shape memory effect in a gold-cadmium alloy.

*1936*; W.B. Elwood invents the magnetic reed switch.

*1938;* Chester Carlson invents the principle of photocopying (xerography).

*1938;* Roy Plunkett accidentally invents a nonstick plastic coating called Teflon.

*1939;* Igor Sikorsky builds the first truly practical helicopter.

*1940s;* English physicists John Randall and Harry Boot develop a compact magnetron for use in airplane radar navigation systems.

*1942;* Enrico Fermi builds the first nuclear chain reactor at the University of Chicago.

*1945;* US government scientist Vannevar Bush proposes a kind of desk-sized memory store called Memex, which has some of the features later incorporated into electronic books and the World Wide Web (WWW).

*1947;* John Bardeen, Walter Brattain, and William Shockley invent the transistor, which allows electronic equipment to make much smaller and leads to the modern computer revolution.

*1949;* Bernard Silver and N. Joseph Woodland patent barcodes—striped patterns that are initially developed for marking products in grocery stores.

*1950s;* Charles Townes and Arthur Schawlow invent the maser (microwave laser). Gordon Gould coins the word "laser" and builds the first optical laser in 1958.

*1950s;* Stanford Ovshinksy develops various technologies that make renewable energy more practical, including practical solar cells and improved rechargeable batteries.

*1950s;* European bus companies experiment with using flywheels as regenerative brakes Flywheels.

*1950s;* Percy Spencer accidentally discovers how to cook with microwaves, inadvertently inventing the microwave oven.

*1954;* Indian physicist Narinder Kapany pioneers fiber optics.

*1956;* First commercial nuclear power is produced at Calder Hall, Cumbria, England.

*1957;* Soviet Union (Russia and her allies) launch the Sputnik space satellite.

*1957;* Lawrence Curtiss, Basil Hirschowitz, and Wilbur Peters build the first fiber-optic gastroscope.

*1958;* Jack Kilby and Robert Noyce, working independently, develop the integrated circuit.

*1959;* IBM and General Motors develop Design Augmented by Computers-1 (DAC-1), the first computer-aided design (CAD) system.

*1960s;* Joseph-Armand Bombardier perfects his Ski-Doo® snowmobile.

*1960;* Theodore Maiman invents the ruby laser.

*1962;* William Armistead and S. Donald Stookey of Corning Glass Works invent light-sensitive (photochromic) glass.

*1963;* Ivan Sutherland develops Sketchpad, one of the first computer-aided design programs.

*1964;* IBM helps to pioneer e-commerce with an airline ticket reservation system called SABRE.

*1965;* Frank Pantridge develops the portable defibrillator for treating cardiac arrest patients.

*1966;* Stephanie Kwolek patents a super-strong plastic called Kevlar.

*1966;* Robert H. Dennard of IBM invents dynamic random access memory (DRAM).

*1967;* Japanese company Noritake invents the vacuum fluorescent display (VFD). Vacuum fluorescent displays.

*1968;* Alfred Y. Cho and John R. Arthur, Jr invent a precise way of making single crystals called molecular beam epitaxy

(MBE).

*1969;* World's first solar power station opened in France.

*1969;* Long before computers become portable, Alan Kay imagines building an electronic book, which he nicknames the Dynabook.

*1969;* Willard S. Boyle and George E. Smith invent the CCD (charge-coupled device): the light-sensitive chip used in digital cameras, webcams, and other modern optical equipment.

*1969;* Astronauts walk on the Moon. Space rockets

*1960s;* Douglas Engelbart develops the computer mouse.

*1960s;* James Russell invents compact discs.

*1970;* Electronic ink is pioneered by Nick Sheridon at Xerox PARC.

*1971;* Ted Hoff builds the first single-chip computer or microprocessor.

*1973;* Martin Cooper develops the first handheld cellphone (mobile phone).

*1973;* Robert Metcalfe figures out a simple way of linking computers together that he names Ethernet. Most computers hooked up to the Internet now use it.

*1974;* First grocery-store purchase of an item coded with a barcode, Barcodes and barcode scanners.

*1975;* Whitfield Diffie and Martin Hellman invent public-key cryptography.

*1975;* Pico Electronics develops X-10 home automation system, Smart homes.

*1976;* Steve Wozniak and Steve Jobs launch the Apple I: one of the world's first personal home computers.

*1970s– 1980s;* James Dyson invents the bagless, cyclonic vacuum cleaner.

*1970s– 1980s;* Scientists including Charles Bennett, Paul Benioff, Richard Feynman, and David Deutsch sketch out how quantum computers might work.

*1980s;* Japanese electrical pioneer Akio Morita develops the Sony Walkman, the first truly portable player for recorded music.

*1981;* Stung by Apple's success, IBM releases its own affordable personal computer (PC).

*1981;* The Space Shuttle makes its maiden voyage.

**1981;** Patricia Bath develops laser eye surgery for removing cataracts.

*1981;* Fujio Masuoka files a patent for flash memory—a type of reusable computer memory that can store information even when the power is off.

*1981– 1982;* Alexei Ekimov and Louis E. Brus (independently) discover quantum dots.

*1983;* Compact discs (CDs) are launched as a new way to store music by the Sony and Philips corporations.

*1987;* Larry Hornbeck, working at Texas Instruments, develops DLP® projection—now used in many projections TV systems.

*1989;* Tim Berners-Lee invents the World Wide Web.

*1990;* German watchmaking company Junghans introduces the

MEGA 1, believed to be the world's first radio controlled wristwatch.

*1991;* Linus Torvalds creates the first version of Linux, a collaboratively written computer operating system.

*1994;* American-born mathematician John Daugman perfects the mathematics that make iris scanning systems possible.

*1994;* Israeli computer scientists Alon Cohen and Lior Haramaty invent VoIP for sending telephone calls over the Internet.

*1995;* Broadcast.com becomes one of the world's first online radio stations, Streaming media.

*1995;* Pierre Omidyar launches the eBay auction website.

*1996;* WRAL-HD broadcasts the first high-definition television (HDTV) signal in the United States.

*1997;* Electronics companies agree to make Wi-Fi a worldwide standard for wireless Internet.

**21st century**

*2001;* Apple revolutionizes music listening by unveiling its iPod MP3 music player.

*2001;* Richard Palmer develops energy-absorbing D3O plastic.

*2001;* The Wikipedia online encyclopedia is founded by Larry Sanger and Jimmy Wales.

*2001;* Bram Cohen develops BitTorrent file-sharing.

*2001;* Scott White, Nancy Sottos, and colleagues develop self-healing materials.

*2002;* iRobot Corporation releases the first version of its

Roomba® vacuum cleaning robot.

*2004;* Electronic voting plays a major part in a controversial US Presidential Election.

*2004;* Andre Geim and Konstantin Novoselov discover graphene.

*2005;* A pioneering low-cost laptop for developing countries called OLPC is announced by MIT computing pioneer Nicholas Negroponte.

*2007;* Amazon.com launches its Kindle electronic book (e-book) reader.

*2007;* Apple introduces a touchscreen cellphone called the iPhone.

*2010;* Apple releases its touchscreen tablet computer, the iPad.

*2010;* 3D TV starts to become more widely available.

*2013;* Elon Musk announces "hyperloop"—a giant, pneumatic tube transport system.

*2015;* Supercomputers (the world's fastest computers) are now a mere 30 times less powerful than human brains.

*2016;* three nanotechnologists win the Nobel Prize in Chemistry for building miniature machines out of molecules.

*2017;* Quantum computing shows signs of becoming a practical technology.

(b) Present Day Sci-Tech Wonders:

# Recent Discoveries in Science and Technology - Progressive Advancement

The following most recent science and technology discoveries indicate the fast and far reaching level and pace achieved by advance and industrialized countries. This global Scientific and technological advancement (STA) has shown how far Africa and indeed Nigeria is left behind and invariably exposed Nigeria's backwardness in modern science and technology.

1. The United States of America has began planning to create a space defense force after a call in July, 2018 by the president Donald John Trump to the country's apex science and technology body NASA to create a "U.S Space defense force" (USDF) by 2020. It will serve as a new domain for battle and defense of the country's space infrastructures such as the International Space Station, Earth Orbiting Satellites of various purposes, including Radio and communication satellites and protection of space route for the much awaited Space X – space tourism shuttling crafts. The Space defense force will be a branch of the army just like other conventional fighting domains in land, sea, air, undersea etc and as a $6^{th}$ arm of the U.S army this will principally be an advanced asymmetric fighting force with standby force in space.

2. NASA sets Out to 'Touch the Sun' by boldly going where no spacecraft has gone before, the Parker Solar Probe will attempt to solve big mysteries about our nearest star.
The Parker Solar Probe launched at 3:31 a.m. ET on August

12,2018 riding a ULA Delta IV rocket on a path toward solar orbit. Equipped with four instrument suites, the probe is—of course—solar-powered, and will be drinking in the energy of the star it's studying over the course of its seven-year journey. Until now, no telescope has ventured close enough to really study the star at the center of it all. Named after 91-year-old astrophysicist Eugene Parker, who first identified the supersonic stream of particles called the solar wind, the probe's science objectives are broadly three-pronged. Billowing out from the sun, the solar wind stretches right to the edge of the solar system, accelerating from a relatively lazy breeze near the star to a faster-than-sound barrage of energy and matter that whips through space at millions of miles an hour "Most of the instruments sit on the main body of the spacecraft and are well in the shadow provided by the heat shield, NASA's historic Parker Solar Probe mission will revolutionize our understanding of the Sun, where changing conditions can propagate out into the solar system, affecting Earth and other worlds. Parker Solar Probe will travel through the Sun's atmosphere, closer to the surface than any spacecraft before it, facing brutal heat and radiation conditions — and ultimately providing humanity with the closest-ever observations of a star. In order to unlock the mysteries of the Sun's atmosphere, Parker Solar Probe will use Venus' gravity during seven flybys over nearly seven years to gradually bring its orbit closer to the Sun. The spacecraft will fly through the

Sun's atmosphere as close as 3.8 million miles to our star's surface, well within the orbit of Mercury and more than seven times closer than any spacecraft has come before. (Earth's average distance to the Sun is 93 million miles.) To perform these unprecedented investigations, the spacecraft and instruments will be protected from the Sun's heat by a 4.5-inch-thick (11.43 cm) carbon-composite shield, which will need to withstand temperatures outside the spacecraft that reach nearly 2,500 F (1,377 C). The primary science goals for the mission are to trace how energy and heat move through the solar corona and to explore what accelerates the solar wind as well as solar energetic particles. Parker Solar Probe is part of NASA's Living with a Star program to explore aspects of the Sun-Earth system that directly affect life and society.

3. Quantum Radar Could Make Stealth Technology Obsolete. Using entangled photons, scientists are on the process of creating 'quantum radar' that can detect stealth bombers to give an early warning system before any attack could be carried out.

4. The Amazing Tech in 'Black Panther' Is More Realistic than is perceived. Much of the tech that is part of everyday life in Wakanda is grounded in technologies that are used today.

5. Wormholes Could Cast Weird Shadows That Could Be Seen by Telescopes. Wormholes could leave a signature swooshed shadow that future telescopes could detect.

6. 'Everything Repellent' is a new clear coating that can keep fingerprints and jelly smears off walls, windows and even

phone screens.

7. A City-Size 'Telescope' Could Watch Space-Time Ripple 1 Million Times a Year. This gravitational wave detector Telescope that will be 25-mile-long is being planned in advance stage by scientist.

8. The world's largest floating solar power plant was completed and connected to the local power grid in China's Anhui province in May 2015. This 40-megawatt solar facility is built on top of a flooded coal mining region. It is part of a vast shift in China's use of fossil fuels. China increased its solar power output by 80 percent in the first three months of 2017, China also completed the Longyangxia Dam Solar Park, a 10-square-mile, land-based solar power plant in 2015. It is allegedly the largest solar facility on the planet.

9. This Quantum Random Number Generator Can Never Be Hacked. A special experimental setup produces certifiably random numbers to use in the creation of "unhackable" messages.

10. Elon Musk the founder of Tesla Industries Worries That AI Research Will creates an 'Immortal Dictator' "It would live forever... and we could never escape," Musk says in a new documentary.

11. An App will detect and Know when an Earthquake will hit any place before it happens. Example, the App knew that an earthquake is Hitting California Before it happens. That's how some people in Los Angeles knew about today's earthquake

before it even hit.

12. South Korean scientists had a lab for creating a 'Killer Robot'. The artificial intelligence (AI) community experts have a clear message for researchers in South Korea: Don't make killer robots.

13. The Lovable 'Star Wars' Droid Creators of the endearing "Star Wars" droid 7 versions of BB-8 revealed how they constructed this adorable mechanical marvel.

14. Google Company has a Military Partnership to produce a smarter military drones.

15. 'Stingray' Spy Devices are state of the art miniaturize devices used to eavesdrop conversations. In Washington D.C alone, Government officials admitted that rogue spying devices are being used in to intercept people's cell phone data.

16. NASA's Supersonic X-Plane Will Tear Through the Sound Barrier with 'a Gentle Thump'. The sleek, single-pilot plane will reduce the roaring sonic boom of supersonic travel to a quiet thump.

17. NASA Has a Plan to Put Robot Bees on Mars. These Mars bees would flap their way around the Red Planet, mapping the terrain and collecting air samples.

18. Scientists have devised a fascinating and beautiful way to create watery sculptures within other liquids. This mesmerizing 'Self-Healing' Liquid Sculptures hold their shape and can heal perfectly well.

19. New Mars Model Details the Violent Birth of Phobos and

Deimos. Southwest Research Institute scientists posit a violent birth of the tiny Martian moons Phobos and Deimos, but on a much smaller scale than the giant moon.

20. 'Desiccation Cracks' Provide evidence of Water on Mars, as Curiosity rover marches across Mars, the red planet's watery past comes into clearer focus. In early 2017 scientists announced the discovery of possible desiccation cracks signaling a major breakthrough in the quest for possible existence of water on the red Martian planet.

21. Scientists reveal how Diamonds can bend and Stretch. Diamond is well-known as the strongest of all natural materials and with that strength lays another tightly linked property: brittleness.

22. Integrating Photonics with Silicon nano electronics into Chip designs. This new technique would allow addition of optical communication components to existing chips with little modification of their designs. Two and a half years ago, this technology looks nearly impossible to achieve.

23. Astronomers studying the motions of galaxies and the character of the cosmic microwave background radiation came to realize in the last century that most of the dark matter composed of Primordial Black Holes.

24. NASA scientists are hard at work trying to unlock mysteries of our planet's ocean surface currents and winds using a new Earth science radar instrument. This also corresponds with DopplerScatt Doubles Scientists' view of Ocean-Air

Interactions.

25. Scientists Use 'Spider Silk' for new biodegradable Bone-Fixing Composite. UConn researchers have created a biodegradable composite made of silk fibers that can be used to repair broken load-bearing bones without the complications sometimes presented with present existing machines.

26. Homemade Microscope Shows How a Cancer-Causing Virus Clings to Our DNA. Using a homemade, high-tech microscopes, scientists at the School of Medicine have revealed how a cancer-causing virus anchors itself to our DNA.

27. Super solar expressway that charges electric vehicles as they drive. China is planning to build a solar expressway for self-driving cars and electric vehicles that will be able to charge them as they drive, according to the Chinese newspaper Hangzhou Daily. Sputnik reports that the highway will be 161 kilometers in length between Hangzhou in the east of the country to the port and industrial hub of Ningbo just south east. The lanes will be embedded with photovoltaic cells that the designers say will power cars as they drive, including self-driving vehicles. According to the Global Times, the highway will also be fitted with sensing and monitoring equipment to reduce traffic congestion.

28. In September of this year (2018), the Five hundred meter Aperture Spherical Telescope – FAST for short – is set to open its doors and become operational. First proposed in the early 90s, it will become the biggest single-aperture radio telescope

on the planet, with 4,600 triangular panels. The telescope is situated in a natural basin in Pingtang County, Guizhou Province, to protect the project from unwanted magnetic disruptions.

29. Mobile chip giant Qualcomm will begin to make server chips specifically designed for the Chinese market this year, through a business owned by the Chinese government. The Guizhou province-Qualcomm collaboration was initiated because server demand in the country is expected to eventually outpace that of the US.

30. The Chinese Academy of Sciences' head scientist, Pan Jianwei, recently announced that the country will undertake its first experiments with a 'quantum satellite'– to establish a quantum communications link between earth and space.

31. Although 3D printing is by no means new (nor is it emerging in China alone), in 2014 a Chinese company called WinSun Decoration Design Engineering managed to create a 10-house 3D-printed village in under one day. After printing out each of the prefabricated modules, the components were lifted into place by a crane and were then ready to use. And in 2015, the same company created the world's tallest 3D-printed building at the time.

32. A remote command could one day send immune cells on a rampage against a malignant tumor. The ability to mobilize, from outside the body, targeted cancer immunotherapy inside the body has taken a step closer to becoming reality.

33. New record on squeezing light to one atom: Atomic Lego guides light below one nanometer. All electronic devices consist of billions of transistors, the key building block invented in Bell Labs in the late 1940s. Early transistors were as large as one centimeter, but now measure about 14 nanometers.

34. Scientists create gold nano particles in water as an experiment that, by design, was not supposed to turn up anything of note instead produced a "bewildering" surprise, according to the Stanford scientists who made the discovery: it's a new way of creating gold nano particles.

35. Electrochemical tuning of single layer materials relies on defects Perfection is not everything, according to an international team of researchers whose 2-D materials study shows that defects can enhance a material's physical, electrochemical, magnetic, energy and catalytic properties.

36. New cancer monitoring technology worth its weight in gold is now a reality. The new blood test using gold nano particles could soon give oncologists an early and more accurate prognosis of how cancer treatment is progressing and help guide the on-going therapy of patients.

37. A Robot was developed for automated assembly of designer nano materials.

38. A new research by Scientist has discovered that Van der Waals heterostructures are assemblies of atomically thin two-dimensional (2-D) crystalline materials that display attractive conduction properties for use in advanced electronic devices.

39. Scientists show how salt lowers reaction temperatures to make novel materials. A dash of salt can simplify the creation of two-dimensional materials, and thanks to Rice University scientists, the reason is becoming clear.

40. Scientists discovered that a 2-D nano sheet expands like a Grow Monster. Grow Monsters are expandable water toys. Whatever you call them, they're plastic-like figurines that swell when placed in water.

41. In the past 12 years, the cost of sequencing human DNA has fallen to one one-millionth of its previous level. This reduction in cost means that the next decade should be a time of "amazing advances in understanding the genetic basis of disease, with especially powerful implications for cancer."

## (c) Some visions of the Future:

## Future Sci-Tech Innovations – We'll soon be using that will possibly transform Humanity.

Elon Musk is perhaps this era's most ambitious innovator. He simultaneously heads a company building rocket ships, SpaceX; another making a popular electric car, Tesla; and another that is a leading provider of solar power, Solar City. When asked what innovation he hoped to live long enough to see but feared he might not, he said, "Sustainable human settlements on Mars." The proceeding content is indicative of the soon expected hi-tech inventions in science and technology that may come into effect, hopefully in 12 to 15 years from now. People have been trying to predict the future since Nostradamus was a lad. However futurologist Ian Pearson has listed some hi-tech innovations that he claims will be surefire hits by 2030.

1. **Dream linking;** Using pillows with conducting fibres in the fabric, it will be possible to see monitor electrical activity from the brain. This will not only show when someone is dreaming, but recent developments indicate that we'll also be able to tell what they are dreaming about. It is also possible (with prior agreement presumably, and when both people are in a dream state at the same time) for two people to share dreams. One could try to steer a friend's dream in the same direction, so that they could effectively share a dream, and may even be able to interact in it.

2. **Shared consciousness;** many people believe we will one day have full links between their brains and an external computer. We will be able to directly access more information outside the brain, making us much smarter, with thought access to most of human knowledge. The link will also allow us to share ideas directly with other people, effectively sharing their consciousness, memories, experiences. This will create a whole new level of intimacy, and let you explore other people's creativity directly. This could certainly be one of the most fun bits of the future as long as we take suitable precautions.

3. **Active contact lenses;** these nifty gadgets will sit in your eyes like normal contact lenses. But they will have three tiny

lasers and a micro mirror to beam pictures directly onto the retina, creating images in as high resolution as your eye can see. This could make all other forms of display superfluous. There is no need to wear a wrist watch, have a mobile phone, tablet or TV but you could still have them visually. The contact lens can deliver a full 3D, totally immersive perfect resolution experience. They will even let you watch movies or read your messages without opening your eyes.

4. **Immortality and body sharing;** While computers get smarter, the brain-IT link will also get better, so you'll use external IT more, until most of your mind is outside your brain. When your body dies, you'll only lose the bits still based in the brain. Most of your mind will carry on. You'll go to your funeral, buy an android body and carry on. Death won't be a career problem. If you don't want to use an android, maybe you'll link into your friends' bodies and share them, just as students hang out on friends' sofas. Life really begins after death.

5. **Smart yoghurt;** a 'quad core' PC has four (4) processors all sharing the same chip, instead of the single one there used to be. This will increase until computers have millions of processors. These might be suspended in gel to keep them cool and allow them to be wired together via light beams. In separate developments, bacteria are being genetically modified to let them make electronic components. Putting this together, smart yoghurt could be the basis of future computing. With potentially

vastly superhuman intelligence, one day your best friend could be a yogurt.

6. **Video tattoos;** it will soon be possible to have electronic displays printed on thin plastic membranes, just like the ones you use for temporary tattoos that you put on your skin. With them you could turn your whole forearm into a computer display. Anyone with ordinary tattoos will wish they'd waited a while. You will also be able to get electronic makeup. You would just wipe it all over your face and then touch it to, and it will instantly become whatever you want. You will be able to change your appearance several times a day depending on your mood.

7. **Augmented reality;** you've seen films where the hero sees the world with computer generated graphics or data superimposed on their field of view. That technology area is developing very fast now and soon we will all be wearing a lightweight visor as we walk around. As well as all the stuff your phone does, it will allow you to place anything you want straight right in front of you. The streets can be full of cartoon characters, aliens or zombies. You can change how people look too, replacing them with your favorite models if you wish.

8. **Exoskeletons;** Polymer gel muscles will be five times stronger than natural ones, so you could buy clothing that gives you superhuman strength. They are too expensive to make today, but not in the future. Imagine free-running and leaping between buildings like a superhero, and having built-in reactive

armor to make you bulletproof too, with extra super-senses also built in. A lot of that stuff is feasible, so exoskeletons might become very popular leisure and sportswear, as well as the obvious military and emergency service uses.

9. **Androids;** Artificial intelligence is likely to make computers that you can talk to just like humans in the near future. These can easily link wirelessly to robots. Robotics technology will use polymer gel muscles too, and a nice silicone covering could make them very human-like, so they can mix easily with humans as servants, colleagues, guards or companions, pretty much what they do in the movie I, Robot, but with a much nicer appearance and probably much smarter.

10. **Active skin;** tiny skin-cell sized electronic capsules blown into the skin would enable us to record nerve signals associated with any sensation. Then you could relive the experience days or years later. From a favorite ski run to the feel of everyday objects, you can replay the full sensory experience. Computer games will become totally immersive too.

## 2.2 Energy transitions (The gradual phasing out of fossil fuel)

The Kyoto accord outlines the determination of advanced and third world countries to initiate eco friendly solutions and to cut-off completely greenhouse gases and in the process do away with fossil fuel which is the major contributor of green house gases.

Since the dawn of humanity people have used renewable sources of energy to survive — wood for cooking and heating, wind and water for milling grain, and solar for lighting fires. A little more than 150 years ago people created the technology to extract energy from the ancient fossilized remains of plants and animals. These super-rich but limited sources of energy (coal, oil, and natural gas) quickly replaced wood, wind, solar, and water as the main sources of fuel. Nonrenewable energy resources, like coal, nuclear, oil, and natural gas, are available in limited supplies. This is usually due to the long time it takes for them to be replenished. Renewable resources are replenished naturally and over relatively short periods of time. The five major renewable energy resources are solar, wind, water (hydro), biomass, and geothermal.

Fossil fuels make up a large portion of today's energy market, although promising new renewable technologies are emerging and gradually taking over the market.

It is estimated that Nigeria's oil deposits will be exhausted by the 3rd quarter of this century. The 4th quarter of the century and beyond will be cataclysmic if the country does not diversify before then. When that is done, Nigeria may be at a cross road except if she had diversified into other sources of income including industrialization and to invest massively in science and technology to avert the caterstrophy that may greet the complete phasing-out of fossil fuel.

Advanced countries has since been investing and channeling

their energy and resource might on renewable energy sources as alternative to fossil fuel. Because they believe and know that soon enough, in no distant future, fossil fuel will became a record in the annals of history. From electronics to gadgets, from industrial robotics to factory production lines, from automobiles to trains, from airplanes to space shuttles, from earth orbiting satellites to international space station all are being designed, programmed and fabricated to operate on renewable energy sources which have far reaching advantages over any type of fossil fuel ever in existence.

Renewable energy sources are limitless, endless and free. Depleting renewable energy source is practically impossible. This will be like obstructing the Planets from their routing natural revolution around the Sun or challenging the galactic order of Solar system architecture.

The worldwide reliance on burning fossil fuels to create energy could be phased out in a decade, according to an article published by a major energy think tank in the UK. Alternative energy sources are renewable and are thought to be "free" energy sources. They all have lower carbon emissions, compared to conventional energy sources. These include Biomass Energy (Biomass is biological material derived from living, or recently living organisms. In the context of biomass for energy this is often used to mean plant based material, but biomass can equally apply to both animal and vegetable derived material), Wind Energy, Solar Energy, Geothermal Energy

(Geothermal energy is the heat from the Earth. It's clean and sustainable. Geothermal heat pumps can be used to heat and cool building. From the limited amount of fossil fuels available to their effects on the environment, there is increased interest in using renewable forms of energy and developing technologies to increase their efficiency. This growing industry calls for a new workforce. Advanced and third world countries are talking about eco friendly solutions and at the same time talking about complete eradication of green house gas emissions.. Analyses of how to get to 100% renewable energy typically look at how future energy sources can supply enough energy to meet a given future demand.

It is believed that the next great energy revolution could take place in a fraction of the time of major changes in the past. But it would take a collaborative, interdisciplinary, multi-scalar effort to get there. And that effort must learn from the trials and tribulations from previous energy systems and technology transitions. Moving from wood to coal in Europe, for example, took between 96 and 160 years, whereas electricity took 47 to 69 years to enter into mainstream use. But this time the future could be different, the scarcity of resources, the threat of climate change and vastly improved technological learning and innovation could greatly accelerate a global shift to a cleaner energy future.

For example, Ontario completed a shift away from coal between 2003 and 2014; a major household energy programme in

Indonesia took just three years to move two-thirds of the population from kerosene stoves to LPG stoves; and France's nuclear power programme saw supply rocket from four per cent of the electricity supply market in 1970 to 40 per cent in 1982. Each of these cases has in common strong government intervention coupled with shifts in consumer behavior, often driven by incentives and pressure from stakeholders."The mainstream view of energy transitions as long, protracted affairs, often taking decades or centuries to occur, is not always supported by the evidence. "Moving to a new, cleaner energy system would require significant shifts in technology, political regulations, tariffs and pricing regimes, and the behavior of users and adopters."Left to evolve by itself – as it has largely been in the past – this can indeed take many decades. A lot of stars have to align all at once. "But we have learnt a sufficient amount from previous transitions that future transformations can happen much more rapidly."

In Science Daily (2009) some scientists in Singapore and Switzerland claim that converting the rubbish that fills the world's landfills into biofuel may be the answer to both the growing energy crisis and to tackling carbon emissions. New research published in Global Change Biology: Bioenergy reveals that replacing gasoline with biofuel from processed waste could cut global carbon emissions by 80%. Their study suggests that fuel from processed waste biomass, such as paper and cardboard, is a promising clean energy solution. If developed fully, biofuel could simultaneously meet part of the world's energy needs and also combat carbon emissions and fossil fuel dependency. The study found that 82.93 billion liters of cellulosic ethanol could be produced from the world's landfill waste and that by substituting gasoline with the resulting biofuel, global carbon emissions could be cut by figures ranging from 29.2% to 86.1% for every unit of energy produced. From Garbage Electricity CHAMCO USA, city wastes, which are costly to dispose, are handled in the most environmental friendly manner on the site to generate electricity and produce fertilizer and construction material. This unit uses Municipal Solid Waste (MSW), even garbage with as low as 800 Kcal/Kg and moisture content of up to 50%. The raw garbage is turned into RDF (Refused Derived Fuel) pellet. The pellets together with 30% auxiliary fuels such as: gas, coal, rice husk, etc., using specially designed boiler, produces heat for steam turbine to run electric generator. In Nigeria there are several garbage dump sites all over; these could serve as source to produce Biomass energy with cheap Science and technology in Nigeria.

Wind power is another rapidly developing alternative to fossil fuel. Societies have taken advantage of wind power for thousands of years. The first known use was in 5000 BC when people used sails to navigate the Nile River. Persians had already been using windmills for 400 years by 900 AD in order to pump water and grind grain. Windmills may have even been developed in China before 1AD, but the earliest written documentation comes from 1219. Cretans were using "literally hundreds of sail-rotor windmills to pump water for crops and livestock." Today, people are realizing that wind power "is one of the most promising new energy sources" that can serve as an alternative to fossil fuel-generated electricity. As of 1999, global wind energy capacity topped 10,000 megawatts, which is approximately 16 billion kilowatt-hours of electricity. That's enough to serve over 5 cities the size of Miami, according to the American Wind Energy Association. Germany, the US, Spain, Denmark, and India are among the world's leading nations in the acquisition of wind energy. Interestingly, the instruments are simple Science and technology devices that could be adopted in Nigeria. If adopted in Nigeria, this will give room to rapid industrialization and a dynamic service sector to create jobs to reduce the rising youth unemployment and underemployment in the society. Government must provide the enabling environment like good road, security of lives and property, good remuneration and zero tolerance to corruption. When corruption is eliminated, honesty and hard work will follow. The citizens will obey the law and order.

Sand Petters (2011) is of the opinion that as Nigeria is situated approximately between 4°N and 13°N, she is geographically favorably located to tap unlimited solar energy, the most dependable renewable energy source. It has been estimated that a yearly average of about 2,300 kwh/m$^2$ of solar energy tails on a horizontal surface in Nigeria. In Lagos, the intensity of solar radiation is about 930w/m$^2$ on a clear sunny day. What Nigeria requires is an affordable solar energy technology and cheap appliances.

The ozone layer is being depleted gradually which is resulting to global warming due to hydrocarbons being released into the atmosphere. To have any chance of preventing dangerous climate change, the world needs to reduce greenhouse gas emissions to net zero or even negative by mid-century. Many experts suggest this means we need to completely phase out fossil fuels and replace them with renewable energy sources such as solar and wind.

United Arab Emirate was told some years back that their crude oil deposit will deplete after 50years. The crown prince at that time thought wisely and immediately consulted and at the end came up with a national program to diversify into tourism. They engage in massive economic restructuring which saw them into building Hotels, Skyscrapers, Resorts, shopping malls, Parks and Gardens and other tourist's attractions. Today, UAE to which Abu Dhabi and Dubai belong are the biggest cities become trademarks and center of attraction for tourists, investors and foreign governments to visit and invest.

Nigerian oil reserve, no matter how long will someday finish. If by share luck it does not finish, it will become an endangered species because it will have no value in the open market and if it has no value, it will not drive any demand and if it does not have demand request, it will be forgotten and set aside. This is no fiction but reality since we now have vehicles that run on solar power, airplanes that were tested to run on solar power, manufacturing plants that operates on solar and wind energies and also a whole city that was powered by a mega solar power project in Australia and other cities.

The future is now. It is not too late for Nigeria as a blessed country with almost every diverse natural resources both gaseous, liquid, solid and human to look inwards and create its indigenous scientific and technological advancement programme, though not totally independent of already postulated theorems of physical laws(since there is no science that exist in isolation). But Nigeria can build on established theorems and mathematical principles and derivations to modify and recreate to suit its specific model and brand indigenous to it. Suffice it to say, that if Nigeria takes a bold and aggressive scientific and technological advancement programme with determination through a paradigm shift, it will in the not distant future catch up and even surpass some nations who are ahead of it.

Some advance countries measure the amount of carbon content any individual or group emits to its environment and the individual or group is made to pay for certain charged fee equal to the amount of carbon content being emitted. This is another effort of discouraging the use of any energy source that emits $Co^2$ into the atmosphere. The quantity of $CO^2$ being discharged in these advanced country's atmosphere is captured and recorded by special advanced sensors placed at various locations in the streets. That is another area where fossil fuel is being gradually phased out of the global system and being replaced with solar, automated and artificially intelligent systems, that one day will take over and make fossil fuel valueless. This generally serves to discourage people from the purchase and use of vehicles and other equipments that operate on fossil fuel because they could be charged high tariff but to go "green and eco friendly" and rather purchase and use hybrid or purebred vehicles and other equipment that works on renewable energy sources.

## 2.3 The Negative Effects of Overpopulation

## (Challenges of rapid population growth against Standard of Living)

## Population Explosion And Its Demerits

Population Explosion is a term used to describe a massive increase in a population over a relatively short period of time. In particular, it is used to describe the increase in world population particularly the population increase that occurred in Africa during the 20th century: it is estimated that the world population passed the six billion mark in mid-October 1999, having doubled since 1960, and with the latest one billion people added in only 12 years. Recent increases in world population are mainly attributable to high population growth rates in developing countries like in Nigeria. Improved health care and nutrition has meant a reduction in the death rates of these countries but without an accompanying decrease in the birth rates. This population explosion is likely to continue in Nigeria and most African countries until the birth rates in these countries are reduced. A combination of education and improved economic standards is needed in order to change attitudes and customs, but this is likely to be a gradual process. The fact that population is growing fastest in Nigeria and some of other Africa's poorest countries, means that they face huge problems of urban and rural degradation, with limited means of improvement.

## Population And Poverty In Nigeria

Nigeria's consistent slumbering has resulted to rapid unregulated population explosion which was inimical to the country's growth, development and maturity in nationhood. The overarching challenge for Nigeria in the decades to come is massive population growth in a context of widespread poverty and many other social problems. The rate of growth of Nigeria's population was 12% per annum. Nigeria's population growth rate is rising fast at an alarming rate and there was a recent forecast that in 2050 it will became the $3^{rd}$ most populated country in the world. Is this a blessing or a course? That depends on what the country is ready to accept and willing to do about it, but so far, population growth has not equated to prosperity.

This may not be a good omen for the country at the moment because rapid growth of population may well make it more difficult for Nigerian government to provide education, health, housing, and employment opportunities for all who need them. Overpopulation will equally make it very difficult if not impossible for Nigerian government to untangle itself and mobilize to industrialize the country away from providing basic infrastructure to the majority citizens of the country who are poor. Example, presently huge sums of money are being rolled out to fund basic infrastructure such as housing construction and renovation through Federal government staff housing loan board (FGSHLB) and the federal Modgage bank (FMB). These monies should have been invested by channeling them to fund industrialization which at the long run would have provided faster, easier and better means of providing these infrastructures.

However, at the same time, a failure by successive Nigerian governments to invest in the social sector and to create a positive climate for fair and equitable economic development has perpetuated the conditions under which couples continue to have large families. The situation is made critical by the fact that 95 per cent of the annual increase in world population takes place in developing countries including Nigeria which is generally least able to provide for them, while the numbers of the world's poorest 48 countries could triple by 2050, according to UN estimates.

Numerous studies have shown that where women have access to education, and especially secondary education, they are more likely to marry later and have fewer children. The same is true in relation to health care. Where women have access to good health services, including those related to reproduction and birth control, they are less likely to have unwanted pregnancies and more likely to have healthy children—thus reducing the incentive to have more babies in case some of them die. The same may be said for other aspects of human development. Where couples do not have to rely on their children for security in old age, and where women have job opportunities and status other than through childbearing, they are more likely to opt for a smaller family.

## Nigeria's Population Versus Standard Of Living;

The rate of rapid unregulated population increase in Nigeria as against Production capacity is a serious threat to Per capita income and Standard of Living enjoyed by each citizen of the country, especially when the population has overshadowed the production capacity per head.

Standard of Living, in economics, refers to an assessment of the level of wealth and prosperity at which people live. Usually it takes into account only material items such as income or ownership of consumer goods. If GDP grows at a higher rate than the population, then standards of living are said to be rising. If the population is growing at a higher rate than GDP, living standards are said to be falling.

Gross Domestic Product (GDP), means the total market value of a country's output of goods and services that are exchanged for money or traded in a market system over a certain period (usually a year or a quarter), regardless of who owns the productive assets. GDP measures the value of all economic activity within a country's borders. GDP is equal to private consumption, plus investment, plus government expenditure, plus changes in stocks, plus exports minus imports. These are known as the components of GDP. The origins of GDP are the proportions contributed by different sectors, for example, agriculture, industry (including manufacturing), and services. In most developed countries services account for between 60 and 70 per cent of GDP, industry for between 25 and 40 per cent, and agriculture for less than 5 per cent. Activities such as child care by parents and household repairs carried out by residents do not form part of GDP. GDP is usually valued at market prices; however, by subtracting indirect taxes and adding subsidies, it can be calculated at factor cost, which gives a more accurate view of the income attributable to factors of production. It can be expressed in constant prices (which is usual) or in current prices (which take inflation into account). GDP may be measured in three ways: by adding up the value of all goods and services produced; by adding up the expenditure on goods and services at the time of sale; or by adding up producers' incomes from the sale of goods or services. In theory, each of these three methods should produce the same result, as output equals expenditure which equals income. However, it is impossible to measure GDP precisely, not least because every country has a black (unofficial) economy. Average national income is one way of assessing living standards and is conventionally arrived at by dividing gross domestic product (GDP) by the population to arrive at a figure for GDP per head. Another alternative indicator of measuring good standard of living is GDP per head in Purchasing Power Parities (PPPs), which take into account how many goods and services can be bought for the GDP per head in local currency. PPP estimates are normally shown on a scale of 0 to 100, with the United States as 100. The differences between GDP per head and GDP per head in PPPs are sometimes negligible and sometimes substantial. Another measure of living standards is the Human Development Index (HDI). First published by the United Nations Development Programme in 1990, this takes GDP per head plus adult literacy and life expectancy into account, thus reflecting to a limited extent the quality of life. Like PPPs the HDI uses a scale of 0 to 100. Judged by it, living standards in Australia, the United Kingdom, Japan, and the United States are all within a two-point range and are among the ten highest in the world.

## Learning From Advanced Countries Population Policies

European countries did not address the issue of a national population policy until the 20th century. Indeed many still do not have an official "population policy". Nevertheless, subsidies were granted to expanding families by such disparate nations as Great Britain, Sweden, and the former USSR. The Italian Fascists in the 1920s and the National Socialists (Nazis) in Germany during the 1930s made population growth an essential part of their supremacist doctrines. Japan, with an economy comparable to those of the European nations, was the first developed country in modern times to initiate a birth-control programme. In 1948 the Japanese government formally instituted a policy using both contraception and abortion to limit family size, while resisting use of the contraceptive pill on health grounds. European pronatalist policies were conspicuously unsuccessful in the 1930s, and their milder variations over the past few decades (in, for example, France and many Eastern European nations) have apparently done little to slow a continuing fertility decline. Sweden has had greater success in increasing family size, with social policies that assist women, and generous family allowances.

## Nigeria and Developing Countries need to Adopt Sustainable Population Policy.

In 1952 India took the lead among developing nations in

adopting an official policy to slow its population growth. India's stated purpose was to facilitate social and economic development by reducing the burden of a young and rapidly growing population. Surveys to ascertain contraceptive knowledge, attitude, and practice showed a high proportion of couples wishing to have no more children. Few, however, practiced efficient contraception. Family-planning programmes were seen as a way to satisfy a desire for contraception by a large segment of the population and also to confer health benefits by spacing and limiting births. However, the use of forcible sterilization and the imposition of family-planning targets caused much resentment—practices which were largely abandoned after the Cairo Conference. In August 1999 India's population reached 1 billion.

Much of East Asia's lowered growth rate can be attributed mainly to the stringent population policies of China. Although it has a huge population, China has reduced both fertility and mortality. The government's one-child per family policy has been credited with averting some 300 million births since it was introduced in the early 1980s, though efforts to stem population growth began a decade earlier. The policy is applied with varying degrees of severity, with greater flexibility in the rural areas and among minority peoples. And, in co-operation with UNFPA, the Chinese government is adopting a new "people-centered" approach to reproductive health in selected counties, under which parents are supplied with high-quality health care and a wide choice of family planning methods, in the belief that this will encourage them to opt for no more than two children. So far, it appears to be working. The government's aim is to limit China's population to 1.4 billion in 2010 and 1.56 billion in 2020. China realized then that the only way to overcome population explosion is to rise above their growing population. This, they did through sustainable policy aided by conscious investment in scientific and technological advancement. By so doing, the production level has rivaled the population increase and the "per capital income" has significantly improved. The unemployment rate has decreased and the economy has been witnessing 4 – 6% annual growth. The economic growth has also translated into better quality of life of its citizens by raising their standard of living. It also improved the physical infrastructural development of the country as can be seen in the number of skyscrapers dotting the skyline in most cities in China, not to mention massive roads, bridges, buildings and water projects all around the country. China is also competing with the most advanced countries of the world in the number of Billionaires they parade despite the fact that it is a socialist developing economy, which means the business investment window through human capital development is growing rapidly. Other Asian countries with specific policies to reduce their levels of fertility or stabilize numbers include Bangladesh, which aims to reach replacement-level fertility by 2005; Nepal, which aims to do so by 2017; and Pakistan, whose aim to reduce its rate of population growth to 1.9 per cent by 2004 has already been overtaken by events, since its rate of growth in that year was 2.7 per cent. Many African countries now have population policies, including the most populous country: Nigeria. With a population of 189 million, Nigeria has seen its numbers multiply six times in the past 80 years. City numbers have grown especially fast, with some 70 per cent of the urban population living below the official poverty line. The country aims to improve the standard of life for all Nigerians by holding the population growth rate to 2 per cent by 2015.

There is now a wide consensus that the type of approach to population policy mapped out at Cairo can make a big contribution to sustainable development. But it will take a more determined effort from developing countries and donor governments before the explosion in human numbers can be safely stabilized.

At present, the per capita income of Nigeria is very poor; it is related to that of poor countries of sub – Saharan Africa. Some countries with lower population than Nigeria even enjoys higher per capita income than Nigeria, this is because when the population of a country is divided by the production per person, it gives the per capita income which in turn defines the standard of living indices a country enjoys. Nigeria with its present massive population estimated at 189 million people, is still a consuming economy with high import dependence and very low production output per head.

Nigeria can rise above its population challenge by adopting the restructuring pattern adopted by Russia, China, India, Mexico and Brazil etc in the past to cushion the effect of rapid population expansion and its resulting demerits. At present, Nigeria imports practically almost everything it consumes. The ratio of exports to imports is 90% in favor of the latter while the former is less than 10% which is mostly crude oil. A slumbering Nigeria, with a rapidly growing population and a weak economy which practically depends on imports rather than exports and an army of unemployed majority is like a timed bomb in the theory of doomsday apocalypse.

## Possible African Population Outlook A Century From Now- What It Connotes To Nigeria

Africa's population looks set to quadruple over the twenty-first century, Africa's population is booming. By 2100, it will be home to 4.4 billion people - four times its current population. Such an increase - far larger than the global population increases of 53 per cent by 2100. For Africa and indeed Nigeria being the most populous country in Africa, this will connote significant challenges. Poverty, conflict, disease and access to education are all issues African governments will continue to face, having to build states that can support ever-increasing amounts of people. Can Africa and Nigeria translate its huge population growth into economic development and improved quality of life? By 2050, more than half of Africa's 2.2bn people will live in its rapidly expanding cities. That's the equivalent of the population of China.

The UN has counted 71 African cities with a population higher than 750,000, many of which lack the infrastructure to support large populations. These cities are growing at an unstoppable pace - expected to hold 100m more people in 2025 than they did in 2010.

Just as Africa and Nigerian governments will have to cope with higher populations, greater levels of urbanization will also challenge countries as they seek to develop. The UN has warned: "The continent continues to suffer under very rapid urban growth accompanied by massive urban poverty and many other social problems.

"These seem to indicate that the development trajectories followed by African nations since post-independence may not be able to deliver on the aspirations of broad based human development and prosperity for all."

## 2.4 Overtaken by previous equals (How they left us slumbering)

As Isaac Asimov has said, "Our world is now future-oriented, you see, in the sense that the rate of change has become so rapid that we can no longer wait until a problem is upon us to work out the solution. If we do, then there is no real solution, for by the time one has been worked out and applied, change has progressed still further and our solution no longer makes sense at all. The change must be anticipated before it happens." Babajide Kolade Otutojo the head of news TVC channels in an interview on the $7^{th}$ of July 2018 said and Quote "A situation where Nigeria is only better than Republic of Congo, Sudan, Niger, Myanmar, is something that is a shame to me ". By failing to move forward, Nigeria inevitably moved backward relative to its rivals and to the environmental and economic threats it faced.

In the 1950s, many foreign observers regarded the Korean economy as hopeless. Until the early 1960s, the economy depended on foreign economic aid, and its per capita income was less than $100, which lagged behind that of many African countries (including Nigeria, Ghana and Kenya), not to mention most Latin American countries. Korea, perhaps with Taiwan, is one of the few countries that grew from poverty to industrial strength comparable to advanced OECD countries. Many experts have attributed the wide gap between Nigerian previous equals (mostly the Asian tiger countries, India and some Latin American countries) and Nigeria to the poverty of leadership. They contended that, with visionary and incorruptible leadership, it is still possible to reposition the socio-political economy because Nigeria has the potential for growth and development.

Recall that in the early sixties and seventies, Nigerian pounds equals the currency value of major super powers and it suppases the currencies of other sister countries by far margin. Recall also that in the mid-seventies to early eighties, Ghanaians are very many in Nigeria and are known to be good English language tutors while many of them are good tailors who are known to move about with their sewing machines on their shoulders. Indians, Pakistanis and Lebanese are also many in Nigeria around that period teaching and operating shops. They all migrated to Nigeria for better life at the period when their countries are facing worst economic challenges. Malaysia and Indonesia, which used to be in the same class with Nigeria in status and development during and shortly after the oil boom has progressed to another class. But today, after so many years, the story is different and is in the reverse. They all left Nigeria and re-organized their countries. Ghanaians have recently announced five years of uninterrupted power supply and are proud and ready to export energy to neighboring countries. Indians are amongst the global fastest developing economy. In addition they are known to possess nuclear weapon capability. They have their indigenous car brand referred to as "TATA". Despite her immense population, India was able to absorb the effect and challenge of its high population with a solution brought about by embracing industrialization aided by scientific and technological advancement. The same story is true for Pakistan which is India's neighbors and long time rivals who themselves possess nuclear weapon capability and are by no means better than Nigeria during the late 50s and early 60s.

Simply put, those countries who use to be Nigerian equals have gained advantage and superiority over her as a result of her slumbering. Nigeria's case is that of both consistent slumbering and complacency at the same time over a long period of time. It is embarrassing, demoralizing and a morale killer for Nigeria to look up to her previous equals for assistance or help despite her various abundant natural resources and skilled manpower also considering the fact that they were all previously presented with a level playing ground. Nigeria must arise from its slumber and position itself amongst the committee of nations and stop grumbling and procrastination. Nigeria's position as a West African regional power broker is grossly inadequate and diminishing her potentials. The country has the capability to be a continental power broker if her potentials is harnessed and positioned rightly. The pain and agony of watching her contemporaries slip past her and becomes overloads (as foreign direct investors, Loan creditors and Grantors of foreign aids) over her or role model to her is too much to bear.

Nigeria must rise above board to catch up and even surpass those countries she started together with on an equal footing, but are today claiming dominance over her. This alone should be a driving force for Nigeria to crave for greatness and stardom, not just in Africa but in the global scene. Now is the time,"once beaten twice shy" is the phrase that should be in the lips of every patriotic Nigerian. Nigeria should channel its anger for losing its birthright to its previous contemporaries towards aggressive pursuit of industrialization aided by investment in scientific and technological advancement. This alone, will catapult or propel the country to catch up, compete and even surpass its former mates; otherwise it will only make the country to retrogress further against its former mates

## Taking Lessons from our previous equals- the Asian tigers

we live in a world of staggering and unprecedented income inequality. Production per person in the wealthiest economy, the United States, is something like 15 times production per person in the poorest economies of Africa and South Asia. Since the end of the European colonial age, in the 1950s and '60s, the economies of South Korea, Singapore, Taiwan and Hong Kong have been transformed from among the very poorest in the world to middle-income societies with a living standard about one-third of America's or higher.

Nigeria and some Asian countries were at the same level of development five decades ago. But why the Asian Tigers are waxing stronger economically and politically, Nigeria's growth appears to have been stalled by its lack of dynamic and transformational leadership. Nigeria and Malaysia have many things in common. They were former British colonies. Malaysia became independent in 1957, three years before Nigeria achieved independence in 1960. The economy of both countries was agrarian, prior to the oil boom era in Nigeria. They are republican states operating bi-camera legislature. Malaysia is a federal democracy leaning on the monarchial order. The Yang di-Pertuan Agong is Head of State and the Prime Minister of Malaysia is the Head of Government. The executive power is exercised by the federal government and the 13 state governments. The federal legislative power is vested in the federal parliament and the state Assembly. The judiciary is independent of the executive and the legislature, though the executive maintains a certain level of influence in the appointment of judges. The Constitution of Malaysia is codified and the system of government is based on the Westminster system, which Nigeria practiced between 1960and 1966. The hierarchy of authority, in accordance to the federal constitution, consists of the executive, judiciary and legislative. The Parliament is made up of the Senate (Upper House) and House of Representatives (Lower House). The country practices a multi-party system, which in the last 50 years, has been dominated by the United Malays National Organization (UMNO), the dominant party in the broad-based coalition called the National Front. Three main opposition parties compete in national and state elections in Malaysia. During the tenure of Dr Mahathir Mohammad, the fourth Prime Minister of Malaysia, many constitutional amendments were made. Henceforth, the Senate can only delay a bill from taking effect and the monarch no longer has veto powers on proposed bills. Unlike several other Islamic societies, Malaysia practices a liberalized form by allowing even western attires to be worn by women. Contact between sexes is fairly flexible. Women account for 40 per cent of the population.

Malaysia has built a "knowledge economy" woven around hi-tech and all-round specialization. Technology and telecommunication are already advanced. Mobile phones were common place, since 2001, even among school children.

Malaysia is cited as a good example of well managed capitalism. Every sector of the economy is tightly controlled by government, with indigenous expertise driving them. For more than 20 years, when Dr Muhammad was the Prime Minister, his Finance Ministers had consistently emerged the second most powerful men, a reward for watching the health of the economy. There are no subsidies in Malaysia. Though government is active in every sector, this is in the form of ensuring compliance with economic policies by both the public and private sectors. At the turn of the last century, large quantity of palm seedlings was ferried from Nigeria for transplant in Malaysia. Today, the country boasts of millions of flourishing palm plantations all over its land scope. Indeed, Malaysia is the world's largest producer of palm oil and third largest producer of rubber.

The country has a national car called Proton, manufactured in collaboration with Mitsubishi of Japan. There are other less popular brands. In 1998, the bubble burst and the country, along its neighbours, plunged into its first recession. It promptly tightened capital controls and rejected prescriptions of the International Monetary Fund (IMF). It reduced the 21 commercial banks, 12 merchant banks and 25 finace houses to only six in each segment. The economy bounced back two years later. Its Gross Domestic Product (GDP) averaged 12 per cent in over two and half decade resulting in a bullish emerging market and powerful member of the famed Asian Tigers. To underscore Malaysia's technological and financial muscles, the nation's capital, Kuala Lumpur currently houses one of the world's tallest office complexes, the Petronas Towers, which is 1,482 feet high, all massive steel and glass. It has 88 storeys and was completed in 1996.

The other four Asian Tigers, are the economies of Hong Kong, Singapore, South Korea and Taiwan, which underwent rapid industrialization and maintained exceptionally high growth rates (in excess of 7 percent a year) between the early 1960s (mid-1950s for Hong Kong) and 1990s. By the early 21st century, all four had developed into high-income economies, specializing in areas of competitive advantage. Hong Kong and Singapore have become world-leading international financial centres, whereas South Korea and Taiwan are world leaders in manufacturing electronic components and devices. Their economic success stories were credited to neoliberal policies with the responsibility for the boom, including maintenance of export-oriented policies, low taxes, and minimal welfare states; institutional analysis also states some state intervention was involved. However, others argued that industrial policy and state intervention had a much greater influence.

The Growth in per capita GDP in the Asian tiger economies (commonly referred to as "the Asian Miracle") between 1960 and 2014 has been attributed to export oriented policies and strong development policies. Unique to these economies were the sustained rapid growth and high levels of equal income distribution. A World Bank report suggests two development policies among others as sources for the Asian miracle: factor accumulation and macroeconomic management.

The Hong Kong economy was the first out of the four to undergo industrialization with the development of a textile industry in the 1950s. By the 1960s, manufacturing in the British colony had expanded and diversified to include clothing, electronics, and plastics for export orientation. Following Singapore's independence from Malaysia, the Economic Development Board formulated and implemented national economic strategies to promote the country's manufacturing sector. Industrial estates were set up and foreign investment was attracted to the country with tax incentives. Meanwhile, Taiwan and South Korea began to industrialize in the mid-1960s with heavy government involvement including initiatives and policies. Both countries pursued export-oriented industrialization as in Hong Kong and Singapore. The four countries were inspired by Japan's evident success, and they collectively pursued the same goal by investing in the same categories: infrastructure and education. They also benefited from foreign trade advantages that sets them apart from other countries, most significantly economic support from the United States; part of this is manifested in the perforation of American electronic products in common households of the Four Tigers. By the end of the 1960s, levels in physical and human capital in the four economies far exceeded other countries at similar levels of development. This subsequently led to a rapid growth in per capita income levels. While high investments were essential to their economic growth, the role of human capital was also important. Education in particular is cited as playing a major role in the Asian economic miracle. The levels of education enrollment in the Four Asian Tigers were higher than predicted given their level of income. By 1965, all four nations had achieved universal primary education. South Korea in particular had achieved a secondary education enrollment rate of 88% by 1987. There was also a notable decrease in the gap between male and female enrollments during the Asian miracle. Overall these advances in education allowed for high levels of literacy and cognitive skills.

The creation of stable macroeconomic environments was the foundation upon which the Asian miracle was built. Each of the four other Asian Tiger states managed, to various degrees of success, three variables in: budget deficits, external debt and exchange rates. Each Tiger nation's budget deficits were kept within the limits of their financial limits, as to not destabilize the macro-economy. South Korea in particular had deficits lower than the OECD average in the 1980s. External debt was non-existent for Hong Kong, Singapore and Taiwan, as they did not borrow from abroad. Although South Korea was the exception to this - its debt to GNP ratio was quite high during the period 1980-1985, it was sustained by the country's high level of exports. Exchange rates in the Four Asian Tiger nations had been changed from long-term fixed rate regimes to fixed-but-adjustable rate regimes with the occasional steep devaluation of managed floating rate regimes. This active exchange rate management allowed the Four Tiger economies to avoid exchange rate appreciation and maintain a stable real exchange rate.

Export policies have been the de facto reason for the rise of these Four Asian Tiger economies. The approach taken has been different among the four nations. Hong Kong and Singapore introduced trade regimes that were neoliberal in nature and encouraged free trade, while South Korea and Taiwan adopted mixed regimes that accommodated their own export industries. In Hong Kong and Singapore, due to small domestic markets, domestic prices were linked to international prices. South Korea and Taiwan introduced export incentives for the traded-goods sector. The governments of Singapore, South Korea and Taiwan also worked to promote specific exporting industries, which were termed as an export push strategy. All these policies helped these four nations to achieve a growth averaging 7.5% each year for three decades and as such they achieved developed country status. The Four Asian

Tigers recovered from the 1997 crisis faster than other countries due to various economic advantages including their high savings rate (except South Korea) and their openness to trade. Gross domestic product (GDP); in 2013, the combined economy of the Four Asian Tigers constituted 3.81% of the world's economy with a total Gross Domestic Product (GDP) of 2,366 billion US dollars. The GDP in Hong Kong, Singapore, South Korea and Taiwan was worth 274.01 billion, 297.94 billion, 1,304.55 billion and 489.21 billion US dollars respectively in 2013, which represented 0.44%, 0.48%, 2.10% and 0.79% of the world economy. Together, their combined economy is close to United Kingdom's GDP of 4.07% of the world's economy.

Education and technology; these four countries invested heavily in their infrastructure as well as in developing the intellectual abilities of their human talent, fostering and retaining their educated population to help further develop and improve their respective countries. This policy turned out to be so effective that by the late 20th century, all four countries had developed into advanced and high-income industrialized developed countries, developing many different areas of advanced technology that give them a tremendous competitive advantage in the world. For example, all four countries have become top level global education centers with Singapore, Taiwan, South Korea and Hong Kong high school students consistently outperforming all other countries in the world and achieving the highest top scores on international math and science exams such as the PISA exam and with Taiwan students winning multiple gold medals every year consistently at the International Biology Olympiad, International Linguistics Olympiad, International Physics Olympiad, International Earth Science Olympiad, International Mathematical Olympiad and International Chemistry Olympiad.

Additionally, these four countries are home to some of the most prestigious top ranking universities in the world such as National Taiwan University, Seoul National University, National University of Singapore, Nanyang Technological University and University of Hong Kong, Faculty of Dentistry, which as of 2017, was ranked as the number one top dental school in the world. While Taiwan and South Korea invested in technological innovation and development, Hong Kong and Singapore pursued a different path of finances and both became world-leading international financial centers. Inspired in part by Japan's technological and economic success, two of the earliest countries to pursue a similar path of cutting edge science and technology development were Taiwan, which has the best and most technologically advanced top ranked medical care system in the world and South Korea, which have both become advanced innovative world leaders in state of the art technologies including medical science, computer technology, biotechnology, space technology (manned spacecraft & robots), military technology, stealth technology, robotics and information technology manufacturing. Both Taiwan and South Korea achieved this by promoting technological innovation, research and development, and export-oriented industrialization which turned an initially post-World War 2 poor agricultural economy into two thriving economic and technological superpowers on the same competitive level as Japan and the United States.

Cultural basis; the role of Confucianism has been used to explain the success of the Four Asian Tigers. This conclusion is similar to the Protestant work ethic theory in the West promoted by German sociologist Max Weber in his book 'The Protestant Ethic and the Spirit of Capitalism'. The culture of Confucianism is said to have been compatible with industrialization because it valued stability, hard work, discipline, and loyalty and respect towards authority figures. There is a significant influence of Confucianism on the corporate and political institutions of the Asian Tigers. Prime Minister of Singapore Lee Kuan Yew advocated Asian values as an alternative to the influence of Western culture in Asia.

Meanwhile in Nigeria, People want to work! Graduates are being pushed out of tertiary institution every day! Ex-NYSC members are always going to every nooks and crannies after their POP in search of white collar job! Most of all these graduates were only exposed to theoretical approach in their higher institutions rather than self-reliant, entrepreneurial and technical approach! Consequentially, this resulted in Nigeria's underdevelopment. If there should be anything like curriculum reform in Nigeria's education system, it should be with respect to the introduction of productive courses and not much humanities. For Nigeria to experience a paradigm shift from being a consuming nation to a producing nation, we MUST learn from the Asian Tigers by adopting their tested solutions that helped them experienced an industrial revolution!!! The replica model of the various tested solutions that worked for them in should be applied to Nigeria; both theoretical and practical approaches are needed at this point, not mere rhetoric and propaganda.

# CHAPTER 3:

## ARRESTING THE VISCIOUS CIRCLE OF NIGERIA'S SLUMBERING

### 3.1 A Call for Nation Building:

For Nigeria to make progress in economic development and nation building it is imperative that her ruling elite decolonize their minds, that is, liberate their minds from foreign control, so they can see and appreciate clearly the country's problems, and mobilize her potentials for effective development.

When Mikhail S. Gorbachev (1931- ) became general secretary of the Communist Party of the Soviet Union in March 1985, he launched his nation on a dramatic new course. His dual program of "perestroika" ("restructuring") and "glasnost" ("openness") introduced profound changes in economic practice, internal affairs and international relations. Within five years, Gorbachev's revolutionary program swept communist governments throughout Eastern Europe from power and brought an end to the Cold War (1945-91. Gorbachev realized that he had inherited significant problems. Even as the USSR vied with the United States for global political and military leadership, its economy was struggling, and its citizens were chafing under their relatively poor standard of living and lack of freedom. Those difficulties were also keenly felt in the Communist nations of Eastern Europe that were aligned with and controlled by the Soviets.

Nehru, the late prime minister of India could not have put it more succinctly when he said: "I do not see any way out of our vicious circle of poverty except by utilizing the new sources of power which science has put at our disposal".

In June 2016, Chinese president Xi Jinping outlined his vision for China to become the leading player in science and technology globally. Speaking at the national congress of the China Association for Science and Technology, he said the country must be on course to being a leading innovator worldwide by 2030. Quote "Great scientific and technological capacity is a must for China to be strong and for people lives to improve," adding that the country and even humankind "won't do without innovation, nor will it does if the innovation is carried out slowly."

Donald John Trump, US's 45th president in his inaugural speech was quoted as saying "America first!! America first!! America first!!

We can equally say Nigeria first! Nigeria first! Nigeria first, Yes we can make it. An internal movement and drive for patriotism as against forces of external influence. We, in Nigeria, have to be more self-reliant. We need to develop innovative ways to diversify our economy, and grow sustainably. It is often said that innovation is the central issue in economic prosperity.

In the words of the famous American novelist, Bill Bryson – "if you think of a single problem confronting the world today – diseases, poverty, global warming, and so on. If the problem is going to be solved, it is science that is going to solve it. If anyone ever cures cancer, it will be a guy with a science degree or a woman with a science degree".

The late Indian iconic freedom fighter Mahatma Gandhi said that "A nation's culture resides in the hearts and in the soul of its people" and Bhagat Singh said that "Bombs and pistols do not make a revolution. The sword of revolution is sharpened on the whetting-stone of ideas".

Still concerning nation building, Maulana Wahiduddin Khan said that "No nation can ever hold up its head, far less take pride of place amongst the nations of the world, if the individuals of whom it is comprised think of nothing but personal gain and self-glorification".

According to Abraham Lincoln the 16$^{th}$ president of the U.S and one of the most greatest said "May our children and our children's children to a thousand generations, continue to enjoy the benefits conferred upon us by a united country, and have cause yet to rejoice under those glorious institutions bequeathed us by Washington".

He also wrote in one of his famous letters that "Adhere to your purpose and you will soon feel as well as you ever did. On the contrary, if you falter, and give up, you will lose the power of keeping any resolution, and will regret it all your life". In another letter he wrote "It teaches that in this country, one can scarcely be so poor, but that, if he will, he can acquire sufficient education to get through the world respectably."

Amongst one of his numerous famous historical speeches, he was quoted as saying "Fellow-citizens, we cannot escape history. We of this Congress and this administration will be remembered in spite of ourselves. No personal significance, or insignificance, can spare one or another of us. The fiery trial through which we pass, will light us down, in honor or dishonor, to the latest generation".

## The Nation Builders

To put it mildly, it is that caring elite so concerned with lifting up that lower rung of the society that qualifies him or her as a nation builder as aptly envisioned in Frantz Fanon's 'Wretched of the earth'. They manifest in various forms: monarchs, aristocrats, autocrats, democrats, plutocrats, technocrats and other types of governance and leadership styles. Nevertheless, the central focus of these leaders is the consuming zeal to make their society better and more fulfilled than they met it. Obviously stepped in their societal role-expectation of a guaranteed future; sundry leaders give what it takes to ensure just that. Even on a universal reckoning, humanity can be distinctly grouped on a three-dimensional classification: the upper, middle and lower cadre(s). While the upper echelon preoccupies itself with its self-defined supremacy, the middle-layer strives to remain relevant in the 'scheme of prevailing intrigues' and the lower rung fights for that oxygenating piece of survival! Regardless of individualities or memberships of each of these societal casings, one thing is predominantly quintessential and that is the relational build-up between and among these different groups. Indeed, no nation is totally insulated against one or another form of discriminations; rather it is the deliberate effort of consensual relationships that makes the whole difference. Some remarkable solid nations had conscientious former leaders who gave all their endowments to nurture their nascent nations along the charted course to full nationhood who their successor-leaders found as guiding relevance afterward. One example is the state of Israel which won a hard independence after May 1948, Led and guided by such visionary leaders as David Ben Guerion, Golder Mear and so on.

Or the case of the United States of America vis-a-vis its struggle for autonomy and self-dictated pace of independence, early in its formative years, from its former old fox-master(Great Britain) and this feat America got with the concerted efforts of its freedom, liberty and civil rights-agitators. Today these combined virtues are what the world at large celebrates as the bastion of democratic principles and representative governance. Either by direct inferences or illustrative practicalities, the league of George Washington, Thomas Jefferson, Abraham Lincoln and John Fitzgerald Kennedy plays the expected role of what contemporary American leaders should be. Nothing less or else. In a positive way, this is one basic and astounding reason why the son of an immigrant settler of a minority group, Barack Hussein Obama, could be so overwhelmingly elected, on two consecutive terms, to the highest office of an almost two-and-half-century republic of over 300 million citizens with hardly any iota of regret or reproach and this is characteristically American!

Or the sacrificial life-style of Africa's renowned icon, Nelson Mandela, who though already gone in the way of all flesh, his memorable landmarks are so conspicuous on the sand of time. Mandela, it was, who elected to spend 27 years of his productive years in torturing imprisonment for the sake of freedom and liberty, equity and impartial justice. As if on a sworn oath of self-immolation, Mandela not only refused but also rejected all offers of freedom from incarceration of about three decades that was short of universal acceptability because sound freedom and accruing rights for one should be the same for another without any layer of ambiguity whatsoever the creed, race, status of such concerned.

In the summit of nation builders, past or contemporary, certain ones have been so outstandingly worthy or disgustingly evil. At the outset of his so-called Aryan race-driven, third Reich in the early 1930s, Adolf Hitler, condemn or commend him it was during his tenure that Germany birthed what the world came around to know as the Volkswagen Beetle which, ab initio, was meant to be a common man's automobile.

Or the strides of communist revolutionist Fidel Castro's Cuba in training medical personnel which the world has grown so used to and this is underlain with its exportation of medical skills and related fields of knowledge.

Or the instructive case of Chairman Mao's China is a lesson indeed of how fundamental and foundational genuine nation builders are to their nations at the outset of any sound developmental process. Chinese revolutionary leader and statesman Mao Zedong and his political and developmental activities that spanned over half a century from the founding of the Chinese Communist party in 1920, to the Long March towards north-west China to escape attack from the nationalist forces of Chiang Kai-shek (1935), from the struggle against the Japanese invaders (1937-1945) and the victory of the Communist revolution (1949), to the Cultural Revolution of the late 1960s, Mao Zedong it was, who initiated the required necessities which modern China now effortlessly prides itself as the world's second largest economy. Before coming this far, China, way back in the late 1940s was shut from the negative influence and corrosive hypotheses of the outside world and having been made to look inward and internalize its bearish strengths be it in sheer population size, diplomacy prominence, economic strength or military capability.

Or the commanding visionary imperial Chancellor, Otto von Bismarck, the architect of German unity, who was the then Prussian Prime Minister and Minister of Foreign Affairs under the Emperor Kaiser William I the king of Prussia, succeeded in uniting the whole of Prussia and other autonomous states to become what is today known as modern federal republic of Germany.

Another nation builder of solid note was Julius Nyerere of Tanzania who gave his all by welding two hitherto discordant nation-states; Tanganyika and Zanzibar and while in charge of the nation-at-large, Nyerere, admirably called the teacher, made deliberate attempts to bequeath a unified people with a common destiny.

Drawing inspirational cues from the above-listed as well as many other well-heeled individuals who shaped their nations; the issue of nation building goes beyond mere wishes of good intents, possession of big pots of money, academic pursuits or other variants for that matter.

For how long can Nigeria afford to trudge on this lonely path of redundancy and self-destruction when on all sides of the globe, men and women of valor are building and shaping their nations from nothing to greatness? The red signals are glaringly suggestive of a sinking abyss? But are there no alternatives, better route to nation building aside the deafeningly and blindingly chosen one by Nigeria's ruling pack since October 1960? Because virtually all directional signs ahead point to one ominous destination: "Road closed to sanity but wide to avoidable self-destruction"!

If anything, nation builders are like the needed catalysts in the developmental processes who from time to time readily explore all weapons in their arsenal to vanquish their opponents. When towards the tail-end of the 20th century, the combined forces of both deceased Ronald Reagan and Margaret Thatcher led the capitalist battle royal against their communist antagonists, their own-defined world was spared the consequential onslaught of the then tagged evil empire.

## Where are the Nation Builders in Nigeria?

Nation-building refers to the process of constructing or structuring a national identity using the power of the state. This process aims at the unification of the people or peoples within the state so that it remains politically stable and viable in the long run. Nation-building can involve the use of propaganda or major infrastructure development to foster social harmony and economic growth (Wikipedia, the free encyclopedia). Originally, nation-building referred to the efforts of newly-independent nations, notably the nations of Africa, to reshape colonial territories that had been carved out by colonial powers without regard to ethnic or other boundaries. These reformed states would then become viable and coherent national entities. Nation-building included the creation of superficial national paraphernalia such as flags, anthems, national days, national stadiums, national airlines, national languages, and national myths. At a deeper level, national identity needed to be deliberately constructed by molding different groups into a nation, especially since colonialism had used divide and rule tactics to maintain its domination.

To understand the notion of nation-building, one needs to have some definition of what a nation is. According to Carolyn Stephenson (2005), early conceptions of nation defined it as a group or race of people who shared history, traditions, and culture, sometimes religion, and usually language. Thus the United Kingdom comprises four nations, the English, Irish, Scottish, and Welsh. The people of a nation generally share a common national identity, and part of nation-building is the building of that common identity. Today the word nation is often used synonymously with state, as in the United Nations. But a state is more properly the governmental apparatus by which a nation rules itself.

For the evolution of nation-building, Almond and Coleman (1960) defined input functions as: He identified equality as one of the basic themes running through all of these. While nation-building after 9/11 still incorporates many of these meanings of political development, equality does not seem to play a major role in practice.

Nation-building that will likely contribute to stable international peace will need to emphasize the democratic participation of people within the nation to demand rights. It will need to build the society, economy, and polity which will meet the basic needs of the people, so that they are not driven by poverty, inequality and unemployment, on the one hand, or by a desire to compete for resources and power either internally or in the international system. This does means not only producing the formal institutions of democracy, but the underlying culture which recognizes respect for the identities and needs of others both within and outside. It means development of human rights-- political, civil, economic and social, and the rule of law. But it also means development of sewer systems, and roads, and jobs. Perhaps most important, it means the development of education. Nation-building must allow the participation of civil society, and develop democratic state institutions that promote welfare. Democratic state-building is an important part of that. This is a multi-faceted process that will proceed differently in each local context.

Many commentators on Nigeria's history and development are always fond of saying Nigeria that is, the country, is an artificial creation of a colonial power, Britain. Let us agree this is true. But is Nigeria the only artificial creation in Africa, or indeed the whole world? Many countries in the world as we have them today are artificial creations. Even the greatest country in the world, The United States of America was not created by God naturally. It was the ability of men of vision and wisdom and sufferings. Most African counties fall into this artificial creation phenomenon. So, why is Nigeria deemed as unique? Is it because we have 250 or so tribes? Is this an insurmountable problem, if indeed it is a problem?

So who builds a nation? Past notable examples of nation builders include Otto von Bismarck (the Iron Chancellor), who united Germany; Kemal Attaturk, who defeated the Ottoman Empire and founded and united present day Turkey. Even, there are the Kwame Nkrumahs, Leopold Senghors, Jomo Kenyattas, Julius Nyereres, Fidel Castros, Mahatma Ghandis of this world. What can be done about nation-building is the question (if it should be done) or who should do it, and who CAN effectively do it. Individual statesmen and women: Where are they in Nigeria? Over the past 50 years, what we have seen are nation-destroyers, not nation-builders. We have been extremely unlucky with our leaders, as well as the followers, at any rate. So, the blame does not lie wholly on the type of leaders our society threw up.

In Nigeria, it has been very difficult to name even one of those people we love to refer to as our Founding Fathers (like the American Pilgrim Fathers) as 'nation building statesmen'. It is really difficult, and this is simply because their mission then was not to build a nation but rather to build power bases and usurp power by whatever means; and mostly serving regional interests.

"The democratic approach to nation building refers to cases in which elected governments operate under inclusive institutions and the leaders behave in ways that strengthen democracy. This approach has the greatest potential for creating a stable multiethnic nation. Unfortunately, Nigerians have not yet successfully pursued this path" (Abu Bakarr, 2004).

Indigenous or exogenous actors: Nation-building is an evolutionary process. It takes a long time. One of the problems with outside actors is that they come and they go, but they are still necessary; arguing for the importance of indigenous nation-building does not mean that outside actors should ignore the process.

As nation builders, let us focus on brain drain of the thousands of graduates leaving the country for greener pastures. This issue of migration has a negative impact on our nation. Nations are build out of human intellect, migration of our many graduates has a serious implication on us. This means that a nation cannot be built without the recognition and the collective efforts of such graduates. (Abiola Saba).

Professor Ibrahim Gambari, in 2006, said "Today, as a nation, we face more challenges than we have known hitherto. Our population has ballooned from 55 million at independence to nearly 130 million. Yet, in our country, children still go to bed hungry and most families subsist on less than one dollar a day. It will, therefore, not be glib to state that in every household, community and state in this nation, where the top hierarchies of human needs are not being met, we certainly have a problem. In a world awash with affluence, yet mired in poverty and hunger we cannot escape our culpability."

This is more so in Nigeria, which once boasted of having

agriculture as its primary industry. Most Nigerians will readily admit that what affects us the most, is poverty and underdevelopment, which are now buffeted by perennial bad governance and debilitating corruption. Likewise, those who are outside Nigeria looking in, will say the same thing, albeit, with a qualifier; to them Nigeria's myriad of problems is self-induced. This often the argument advanced by those who were opposed to any debt forgiveness for Nigeria. They refuse to accept that a nation with so much wealth could be so indigent. To them, our country and the challenges it faces, presents a unique paradox".

Nation-building and the associated developmental issues require men and women of deep vision; sincerity of purpose; selflessness; genuine love for their country and their people; hardworking; of conscience, integrity, credibility, trustworthiness, honesty, reliable and able; people who do not think of stealing or embezzling; people who do not misuse the authority and power conferred on them, by God or Man; people who do not think that getting to positions of authority is a "do-or-die" affair; people who understand the meaning of nation building, leadership, good governance, rule of law, political emancipation, equality, human and civil rights, civility, freedom of speech, rule of law, diversity and religious tolerance,; people who will shun and will not tolerate tribalism, corruption and nepotism. A democratic approach is the best path to nation building in a multi-ethnic country like Nigeria. As we have seen in the Nigerian experience with nation building, it is difficult to pursue a non-democratic means of reform without aggravating internal unrest and international censure. Sadly, the lack of a democratic mandate, poor institutional design, and bad leadership has all made it nearly impossible for successive Nigerian governments to pursue a democratic approach. Many Nigerians are not satisfied with the 1999 Constitution because it failed to address the structural imbalance of the federation (Abu Bakarr, 2004).

## 3.2 Aggressive pursuit of STIR&D

## (Embracing science, technology, Innovations & Research and Development as key to eroding Nigeria's slumber)

Louis Pasteur (1822-1892), the father of Microbiology, knew all along nearly two hundred years ago when he said: "Science knows no country, because knowledge belongs to humanity and is the torch which illuminates the world". George Dyson. Said "I am a technological evolutionist, I view the universe as a phase-space of things that are possible, and we're doing a random walk among them, eventually we are going to fill the space of everything that is possible". Quote;"One of the keys to America's greatness is its ability to make things. As outlined in Manufacturing: A Blueprint for America's Future, industrialization brought our nation to unprecedented levels of prosperity. It carried us to victory in

World War II and took our astronauts safely to the moon and back" as outlined in Building a Nation of Makers.

The word science comes from the Latin "scientia," meaning knowledge. According to Webster's New Collegiate Dictionary (1977), science is "knowledge attained through study or practice," or "knowledge covering general truths about the operation of general laws, especially as obtained and tested through scientific method and concerned with the physical world." According to Wikipedia the free online encyclopedia, the word technology comes from the Greek technología, téchnē means 'craft' and logia is the study of something or the branch of knowledge of a discipline. It is the usage and knowledge of tools, techniques, and crafts or it is systems or methods of organization or a material product. Technologies significantly affect human as well as other animal species' ability to control and adapt to their natural environments.

Science is basically knowledge of how the natural world works. This knowledge is acquired by experimentation and observation in the controlled environment of a laboratory. A scientist is therefore an individual engaged in the business of acquiring knowledge. In contrast, a technologist is involved in applying knowledge to provide practical tasks. A scientist can also be a technologist, and vice versa. The popularity of the term science and technology derives from the fact that human civilization practically depends on it. if one is aware of the power of science and technology in promoting development, one is obligated to continue to implore the Nigerian government and the public to do what most of the advanced nations have done for centuries, and that is, to translate science into productive technologies and use the latter to fuel the development of their country.

Science, technology, research and development are such powerful enterprises that will ensure a nation's development or promote an individual's well-being. So although technology has become very pervasive and can be used to address many problems, a country must pick her technology of choice wisely. Examine the foreign countries from where science and technology was transplanted to Nigeria, we found that these countries have been supporting these enterprises liberally for several hundred years, and that they firmly believe that science and technology is, in part, the basis for their advanced state of development and current high standard of living. Nigeria should emulate this belief in the power of science and technology by supporting and promoting science and technology in our schools.

The only pathway to meaningful industrialization and economic independence is the application of science and technology. The establishment of the technical feasibility of the industrial project; the fabrication of the necessary specific machines and equipment which are peculiar to the target product to be produced, including spare parts, for the factory; the preservation and storage processes for the industrial products; all derive from scientific research and development. Thus, science and technology constitute the intellectual tool for the creation of industrial wealth, and scientists and technologists are the human agents for the attainment of that objective. Science and technology has a very unique side to it – it is intrinsically linked to most, if not all, sectors of an economy so naturally, economic development should be the result of advancing Science and technology. Perhaps nothing sums up the importance of science and technology better than this quote by British Physicist, Stephen Hawking – "The world has changed far more in the past 100 years than in any century in history. The reason is not political or economic but technological – technologies that flowed directly from advances in basic science…"

Recent analysis shows that economic growth over the period 1950 to 2010 is indebted to the innovations, incentives, and productivity gains arising from technological advancements. It is estimated that about 35 percent of the world's GDP growth from 2000 to 2008 can be explained by productivity derived from technological capability and its enhancement through information technology, human capital development, and so on.

Advances in Science and technology can help to diversify the economy, by improving productivity in sectors like agriculture, while defining new ones. Look also at the relatively recent development of the hybrid engines that harness solar power and batteries (for cars, and more-recently for ocean liners and ships) – one of Science and technology's responses to the challenges of carbon emission and exclusive reliance on fossil fuels like diesel and petrol. Imagine the impact of adopting such technology on our country Nigeria? Our fuel subsidy bill, which cost N2.19 trillion (or about 5 percent of GDP), last year, would be substantially lower given lower PMS consumption, creating the fiscal space to invest in other sectors and diversify our economy.

Today's electronics sector, which is driven by an incessant wave of branching innovations that are generating a constantly proliferating range of products, can also aid economic growth and diversification

We could also see a drop in air pollution levels in this country (which is becoming an issue) since the transport sector, together with the power sector, account for about 10 percent of total carbon emissions in Nigeria. The World Bank estimates that with the right technology, Nigeria can generate up to 10,000 MW over the medium to long-term, through Concentrated Solar Power. We currently generate about 4,000 MW from all sources.

Science and technology also plays a key role in improving the quality of life. For instance, research in healthcare has proven vital to the prevention, diagnosis and treatment of various killer diseases. The American Heart Association recently announced that deaths due to coronary heart disease fell by nearly 40 percent in the USA over the last decade due largely to new treatment inventions. The same applies to HIV/AIDS – one of the top three killers of African youth. In 1996, a 20-year old person in the US with AIDS expected to live for about 3 to 5 years, but now expects to live to be 69 years. Only about a few weeks ago was it announced that – Truvada – an HIV fighting pill, can also be used to prevent the disease, after a three-year study. In Nigeria, preventable or treatable infectious diseases such as malaria, pneumonia, diarrhea, measles and HIV/AIDS still account for more than 70 per cent of the estimated one million under-five deaths in Nigeria. Several of these deaths occur as a result of misdiagnosis, due to the poor state of technology in many hospitals. This is why many Nigerians are going abroad, to countries like Egypt, Cuba and India, for medical services (including diagnosis) spending between $600 million and $1 billion annually, according to our health ministry's estimates.

In the education sector, particularly higher education, there is an emerging paradigm shift in the world today. The recent onset of powerful technologies, including cloud computing and precise online assessment regimes enabled the launch of a number of top-tier university entrants into what is being called the Massively Open Online Course (MOOC) marketplace. World class universities ranging from Harvard, Stanford, MIT, are now providing free, high-quality, rigorously assessed and highly accessible online university level education to the masses. it can provide a level of certification that can develop industry-standard skills, for example in the ICT industry, and actually provide a way out for 80 percent of the 1 million Nigerian youth who do not get into universities each year, due to limited supply of college/university places.

In a nutshell, developing countries cannot hope to prosper in an increasingly competitive global economy and open trading system if they do not build the appropriate science and technology capacity to produce more value-added goods and services. In fact, Science and technology is the dividing line between developed nations and those less developed.

We are underperforming, relative to our abundant human capital. According to NEPAD's African Innovation Outlook (2010), South Africa produced over 86,000 scientific papers – about 37 percent of the total research output of 19 African countries surveyed between 1990 and 2009; Egypt produced nearly 60,000 – about 27 percent of output. Nigeria produced 27,743 papers (or 12 percent of the total output) – about one-third of South Africa's output. But a worrying finding is that the productivity growth of Nigeria's scientific research is the second-lowest of the 19 countries. Even though, our scientists doubled their productivity in the period 2005 to 2009 relative to output between 1990 and 1994, other African countries like Algeria and Uganda saw their productivity increase by a factor of 6.3 and 5.4 respectively. To put things in perspective, countries like Brazil and Malaysia saw productivity rise by a factor of above 100.

Democracy cannot be sustained in a vacuum. The government must embark on a process of appropriate technology transfer. Technology holds the lifeblood of any nation especially in today's market-based economies. For our national productivity to improve and for our products to achieve the level of quality necessary to compete in global markets, we must have a national plan and strategy on technology. Adapting technology to suit their needs and training and motivating their citizens to work within their unique cultural framework. In today's competitive environment, time waits for no one. As Nigeria embarks on a diversified economic program, it is timed to initiate technology policies that should not focus entirely on hardware importation and foreign technology transfer but also on software development and culturally sensitive indigenous technology development. People are the most important asset of any organization and indeed, of any nation. If we cannot train and motivate our people to work with our imported technologies and adapt them to our unique conditions, then technology transfer has failed. Failures of technology transfers in both the public and private sector have dire consequences on the national economy. They lead to lower productivity, unemployment, and lack of competitiveness, low foreign currency generation, and dependence on expatriate services.

As we embark on this road towards reconstruction, we must encourage development of indigenous technology and also start refurbishing our infrastructural base. Most of our roads, schools, industries, national monuments etc. are in need of repair. Because of years of neglect, this effort will require time and commitment but must be sustained to be viable. We must develop a maintenance culture. It is more efficient and less expensive. We are a nation with a lot of resources and we should use our vast resources to develop credible technology policies that will guide us through the next millennium. Similarly, South Africa was able to secure more than 1,000 patents in 2010 alone, according to data obtained from the World Intellectual Property Organization (WIPO); whereas, Nigeria secured only 18 patents in the last 8 years. This is unbelievably low. Egypt and Kenya got 604 and 49 patents, respectively, over the same period. There is no doubt that Nigerian scientists are making progress, sending our own satellite (made and launched in China) into orbit. Equally, some developments in medical science are achieved, such as drugs used in combating sickle cell and other diseases; but a majority of these remain at the formative stages and do not become main stream.

## Current realities and status of science and technology in Nigeria

There are a number of reasons for the poor state of our Science and technology sector. Firstly, we need a better and more coherent national strategy, as the sector remains highly fragmented, lacking effective coordination. Even the existence of our Science and Technology Ministry was intermittent, until the end of the 1990s. There are about 106 core research and quasi-research institutes spread across the Ministries, with each one conducting its research without synergy and harmonization. Some of these institutions have been in existence for more than 30 years, yet there is little to show for their work as Nigeria still relies on research done internationally. If public sector research institutes in other countries can develop major technological advances like the internet and the human genome project, what is wrong with our own?

Secondly, our scientists complain about lack of funding. Its agreed that fast growing economies must invest hugely in Science and technology. China has been growing its R&D expenditure by 20 percent annually, since 1999. China now accounts for 12 percent of global R&D expenditure, spending nearly 5 percent of its budget (or 1.76 percent of GDP) in 2010, on the sector. Let's compare this to Nigeria. Over the past decade, government's Science and technology expenditure has been less than 2 percent of the yearly budget (less than 0.3 percent of GDP per year) - a grossly inadequate figure. South Africa spends 8.5 times more on Research and Development than Nigeria but produces over 70 percent of the drugs manufactured in Africa. What do we produce?

It is evident that science and technological transfer and development is solely dependent on science and technology education in the country, for scientists and technologists are definitely required in the economic infrastructure of the society before any scientific and technological development, and industrialization can occur. Even if students do not further their study of science and technology in tertiary institutions and as a result do not go on to become professional scientists, engineers, and technologists, their experience of science and technology gained from the elementary and secondary levels and first year of their tertiary education will be sufficiently rich and relevant. In today's world, science and technology are undoubtedly vehicles for socio-economic development. Decades ago, developed nations were more or less as undeveloped as developing nations of today. They are now transformed from rural, peasant communities into highly urbanized, industrialized countries through the development of their science and technology. In the process, they became rich and politically powerful. For Nigeria to achieve her age-old goal of crossing the borderline between being a developing country and a developed country, she must develop scientifically and technologically. It should be added that science and technology have become integral parts of the world's culture and to lag behind is to be out of place. However, science and technology have continued to have a largely lowly status in developing or underdeveloped societies including Nigeria. On the cause of the current poor levels of science and technology in the third world countries (including Nigeria), Dr. S.A. Thomas in an article on Chemistry in Britain in 1983 said: "Most third world countries may appear to be looking for salvation in science and technology, they lack the foundation necessary to develop their scientific and technological potentials in real terms. Social attitudes favoring non-scientific endeavors and objects leading to a quest for increased material wealth not justified by increased productivity and optimal utilization of scarce resources have continued to inhibit enhanced scientific activities…there is a need to solve the problem of inadequate research leadership and generate a crop of policy makers sufficiently on scientific matter, if third world nations wish to improve their lot."

This foundation necessary to develop science and technology in the country is obviously education—elementary, secondary, and tertiary education. Science and technology have to be taught and studied systematically and purposefully at all levels of education including, at least, first years of tertiary education for the arts and the humanities. Such scientific literacy will equip them to contribute to Nigeria's development in an increasingly competitive and rapidly changing world. In other words, if all students in this developing country, including those in the arts and humanities as well as sciences, were imbued with the curiosity that characterizes scientists and the competence that characterizes engineers and technologists, all would be in a better position to participate in the solution of the indigenous problems of social and economic development. There is a need, therefore, to imbibe the science culture in every Nigerian so as to have the proper and requisite foundation on which to develop our science and technology which will in turn develop the country. This is why great emphasis should be put on effective science education to help lay the needed foundation starting from the primary to, at least, the first year of the tertiary level for students of humanities. The government must design specific policies on science and technical education which must be implemented and sustained to promote science and technology curricula at each level of education. This, of course, must include increased funds which should be provided and properly utilized in the educational system. When scientifically well informed leadership springs up and scientifically and technologically literate citizens abound who are not all about "a quest for increased material wealth" but ways to contribute to problem solving in the society, the right type of environment will exist and illiteracy and superstitions which are prevalent in the country will be eradicated then all these "social attitudes" which have "inhibited third world countries", as Thomas rightly puts it, would have been erased. Nigeria will then have a chance to improve her lot as she finally sets out for scientific and technological development which will in turn hastily pave the way for national development. Underdevelopment which has continually plagued third world countries such as Nigeria, rich in human and natural resources, will be history.

There is no alternative to scientific and technological development. They are what distinguish the underdeveloped countries from the developed ones. The standard of living, social security, military and political power of the country all depend on the advancement of her science and technology. Entrepreneurial centre should be setup to train and retrain young graduates to pick up one vocation or the other. Loans should be made available for establishment of small scale industries by these young Nigerians. Government should fund research in Science and technology and attach grants to breakthroughs. Government should invest in the non- oil and agricultural sector of the economy to absorb research outcomes. The Nigerian government can turn to the very active private sector to provide these basic products and services. Government should redesign and implement effective monetary and fiscal policies in line with current trend in Science and technology. Incentives should be provided to individuals and organizations to invest in the economy, and encourage proper competition to increase the quantity and quality of goods and services. The most important manpower requirement for a successful national science and technology is the president who is also the chief executive of the country, the president because, as a money intensive enterprise science and technology requires a powerful and influential advocate in government since it is the government that allocates money for science and technology. However, the president has to be genuinely convinced that the use of science and technology is the only sure way to develop the country both socially and economically. The president is also crucial because he appoints the minister that oversees the country's efforts in science and technology. Another manpower requirement for science and technology to thrive in Nigeria is the president of the country's National Academy of Sciences (if any). The importance of the National Academy of Sciences president lies in the fact that he is in a unique position to be able to provide the country's president, irrespective of his background, useful information regarding the status of science and technology in the country because he is likely to have a wealth of ideas on the issue. While the Academy president should seek the ear of the country's president, the latter should grant him a cordial audience as the Academy president could not have arrived at such an important position without knowing something about science and technology for development. The third manpower requirement for a successful science and technology in Nigeria is the people who actually carry out the work that becomes the nation's science and technology – the professors, scientists, engineers, laboratory directors, technologists, other technical and support staff, and the educated laymen. All these individuals have one thing in common. They all need to be adequately trained. Adequate training in science and technology takes a long time and should start early in a child's education.

## 3.3 Industrialization:

## (A practical Long-term Development and distraction from the dangers of slumbering)

Industrialization is the conversion, on a large scale, of raw materials to finished products ready for public consumption or for further use in industry. It involves not only the processing of raw materials into finished products but also the preservation, packaging and storage of the finished products. Industrialization ensures that the finished products are in a quality acceptable to the consuming public. Raw materials for industry basically are agricultural produce or mined ore. Industrial products derive from the interaction of matter and energy: energy acting on matter converts it to product. Thus, mechanization and the provision of abundant and cheap electric power are essential and basic to industrial production.

The principal actor in the industrialization effort is the scientist. The task of industrialization is two-fold, first to create wealth, and next to invest wealth wisely to create more wealth. Science creates wealth. Through research, science develops the technical processes to produce goods; it fabricates the machinery and equipment and the spare parts needed for industrial production; it builds and operates the factory for the manufacture of goods; and develops control mechanisms for effective production and quality assurance. The matter of the profitable investment of wealth is the purview of the economist. Technology is science in practice. Through practice processes the scientist develops and advances home-grown technology, and increases the productive capacity of the population.

In consequence, any leadership that ignores or rejects the direct involvement of the nation's scientific manpower in the process of industrialization has rejected technological advancement and economic independence for the nation.

Courting foreign investors as a primary policy for economic development is not the answer. This is the plight of Nigeria today.

Industrialization has advantages. Because it produces goods on a large, commercial scale, the goods are readily available and relatively cheap to the consuming public. It drives the development of mechanization, leading to the development of advanced technology which raises the productive capacity of the nation. It creates massive wealth through large scale production and sale of goods, and hence raises the standard of living of the population. It keeps the people gainfully employed and the urge toward social crime is reduced and may be ultimately eliminated. The country can speak with confidence and pride among the advanced nations of the world.

For the country's industrialization to achieve the desired goal of economic independence, it must be structured and patterned to achieve two objectives, namely,

1. To challenge and tap the creative powers of the country's

scientists, engineers and technologists who will be required to design, fabricate and operate the factories. That is, the country develops home-grown technology through scientific Research and Development (R&D) as the only way to achieve technological advancement.

2. To create industrial wealth in the form of company taxes or profits from direct sales, that will be used to finance the provision of infrastructures and social services, and provide gainful employment for the citizens of the country. In this regard, Nigeria's oil revenue must be treated as finance capital to industrialize the country.

The Historical Development background of industrialization first examined the perpetual war and an inherent weakness in manpower created a demand, which led to increased demand due to increased escalation of conflicts and this propelled the industrial revolution in England. The fear of the English dominance propelled copycat industrial revolutions on the continent and then all over the world. The funding for industrial revolution came in a dispersed form from the Feudal families of the United Kingdom and primarily in England who together numbered no more than a 100 large families who owned most of the land and thus most of the production.

The first stages of the Industrial Revolution occurred in the middle of the 18th century with the early development of the steam engine and textiles manufacture. At this stage development was largely confined to Britain.

One of the major technological breakthroughs early in the Industrial Revolution was the introduction of steam engines. When textile factories first became mechanized, only water-power was available to operate the machinery; the factory owner was forced to locate the establishment near a water supply, sometimes in an isolated and inconvenient area far from a labor supply. After 1785, when a steam engine was first installed in a cotton factory, steam began to replace water as power for the new machinery. Manufacturers could build factories closer to a labor supply and to markets for the goods produced. The development of the steam locomotive and steamship in the early 19th century made it possible to ship factory-built products to distant markets more rapidly and economically, thus encouraging industrialization

The next phase of the Industrial Revolution gathered pace in the last two decades of the 19th century, with the expansion of electric power, the chemical industry, the motor industry, and assembly-line production. It was during the final decades of the 19th century and the early decades of the 20th that the process of industrialization intensified, in terms of the number of industries and the number of countries involved. Between 1880 and 1930 the broad sweep of industrialization included the manufacture of the motor car and its related equipment such as tyres, radio equipment, aircraft, plastics, and stainless steel, and the appearance of the earliest television service.

As new industries were created with the development of steel manufacture and the expansion of the railway system, development also occurred geographically with the spread of these industries across Europe, and especially in Germany, and across the Atlantic to the United States. By the end of the 19th century new industries were opening up in many areas: for example, the production of telephone equipment and domestic electric lighting. During the Renaissance, the advance of science, contact with the New World, and the development of new trade routes to the Far East stimulated commercial activity and the demand for manufactured goods and thereby promoted industrialization. In Western Europe and particularly in England, during the 16th and 17th centuries, many factories were created to produce such goods as paper, firearms, gunpowder, cast iron, glass, items of clothing, beer, and soap. Although heavy machinery, operated by water-power in some places, was used in a few establishments, the industrial processes were generally carried on by means of hand labor and simple tools. In contrast to modern mechanized plants with assembly lines, the factories were merely large workshops where each laborer functioned independently. Nor were factories the most usual place of production; although some workers used their employer's tools and worked on the premises, most manufacturing was done by workers who were supplied with raw materials, worked in their own homes, returned the finished articles, and were paid for their labor. The factory system began to develop in the late 18th century, when a series of inventions transformed the British textile industry and marked the beginning of the Industrial Revolution. Among the most important of these inventions were the flying shuttle patented (1733) by John Kay, the spinning jenny (1764) of James Hargreaves, the water frame for spinning (1769) of Sir Richard Arkwright, the spinning mule (1779) of Samuel Crompton, and the power loom (1785) of Edmund Cartwright. These inventions mechanized many of the hand processes involved in spinning and weaving, making it possible to produce textiles much more quickly and cheaply. Many of the new machines were too large and costly for them to be used at home, however, and it became necessary to move production into factories.

The Process of industrialization entails a transition from an agricultural to an industrial society, associated with a movement towards higher per capita income and productivity levels. For this to happen, the demand for agricultural products (food and raw materials) has to be satisfied. Empirical estimates show that the demand for agricultural goods exhibits income elasticity (as income rises so does the demand for agricultural products). For viable and sustainable industrialization to take place, the demand for agricultural products has to be met either by an increase in imports or by a rapid growth of domestic agricultural productivity. Hence, agricultural productivity growth is a necessary precondition for modern industrial growth to become self-sustaining. Modern industrialization is often dated as having its origins in the British industrial revolution of the 18th century. Sequences of further industrializations were observed during the 19th and 20th centuries. During the 19th century successful industrializations were observed in a number of northern European economies and in North America. By the end of the 19th century this process encompassed some southern European economies and Japan. During the 20th century, particularly after World War II, the process was observed in a number of economies in East Asia.

## Industrialization Models

ONE; "take-off" model of Walt Whitman Rostow analyzed the British and later industrialization experiences. He argued that for countries to industrialize successfully certain prerequisites were needed—such as a high productivity for the agricultural sector, the existence of functioning markets, and stable government. Once the preconditions were present industrialization could manifest itself in the form of a "take-off"—a brief period of 20 to 30 years in which the process of industrialization is completed. Because countries satisfy these preconditions in different historical time periods, industrializations are spread over time.

TWO; "Catching- up" models of industrialization in recent years economists have attempted to explain the process of industrialization within a framework of "catch-up" growth. In this respect the recent theories have developed from Gerschenkron's perspective, which emphasized the historical conditions faced by the late industrializers. Latecomers in terms of industrialization can imitate technologies already in existence in the leading countries, allowing them to undertake economic development and catch up with the per capita productivity levels of the leaders. The theory of catching up predicts that there should occur a convergence of per capita income levels in poor and rich countries.

THREE; "social capability" model of industrialization states that a necessary condition for catching up is a sufficient degree of social capability: the relatively backward economies must be sufficiently socially advanced to be able to adopt the superior technology of the major industrial countries. If countries do not have high levels of human capital, due to a failure to invest in education, or have unstable political systems, we will find that the potential is not taken up. Hence, the whole world is unlikely to industrialize successfully to achieve comparable productivity levels.

## Industrialization through Industries

Primary industries; these are the industries responsible for the extraction of natural resources. They comprise agriculture, hunting, fisheries, forestry, mining, and quarrying. A distinction is often drawn between those primary industries concerned with renewable resources such as forests and those concerned with non-renewable ones, such as minerals.

Secondary industries engage in the manufacturing and production of goods. The word "secondary" implies that such companies are engaged in the second stage of economic activity. They use the natural resources of the primary industries (and possibly the goods of other secondary industries) to make products. Secondary industries include house-building and the manufacture of clothes, food processing, shoes, luggage, furniture, packaging, chemicals, metal products, machinery, electrical products, electronic products, computers, cars, trains, and aeroplanes. They also include utilities, which provide services such as gas, water, and electricity.

Tertiary industries comprise those companies involved in services, as opposed to those providing an extractive or manufacturing function. The tertiary category includes retailers (of clothes, food, and so on), banks, insurance companies, hotels, restaurants, estate agents, lawyers, doctors, accountants, teachers, golf professionals, and television presenters.

The contrast between primary, secondary, and tertiary industries can also be seen at the country level. Advanced economies such as the United Kingdom have a mix of industries, primary, secondary, and tertiary, but with an overwhelming emphasis on the tertiary sector. In contrast, many poorer nations still depend for their livelihoods on primary industries such as minerals or agriculture.       Since World War II the existing industries in the developed world have become much more sophisticated in their products and their manufacturing processes. Miniaturization is only one aspect of this sophistication. At the same time new technologies have encouraged the creation of many new industries, including jet aircraft, computers and electronics in general, satellites, nuclear power, new composite materials, carbon fiber, robotics, telecommunications, and data-processing equipment.

In the latter half of the 20th century there have been continued developments in manufacturing, but also a shift in employment in the advanced economies away from secondary and towards tertiary activities. Higher levels of wealth in the developed world have encouraged the growth of many service industries, such as retailing, the hotel business, tourism and leisure services, and financial and business services. Some of the newer industries have included information technology, cellular radio, biotechnology, and global finance.

Globalization, or the internationalization of production, technology, enterprise, and exchange, means different things to different people. The development of the Internet has massively extended the ability of households to reach out across the globe in search of new information and opportunities.

The growth in world trade and investment has stimulated the growth of multinational corporations. Some of these companies are now economically more powerful than many countries. Seven influences can be identified as driving globalization. They are: increasing global wealth; the removal of trade barriers; the fall of communism; transport and communication developments; company restructuring; the globalization of financial markets; and cultural convergence, epitomized in the development of worldwide brands.

## Africa, Nigeria and Industrialization

The United States emerged from an agrarian economy into an industrial superpower in the 20th century, through effective application of Science and technology. Victor (2009) opined that while vocational and technical education has continued to thrive in many societies, Nigeria has neglected this aspect of education. Consequently, the society lacks skilled technicians to run the economy. Ekpiwhre (2008) asserts that it is important for us to appreciate the potency of Science and technology to bring about significant changes in our local, state and national lives. Investments in Science and technology always pay off, sometimes immediately but always in the long run. In fact, these countries invested quite heavily in people and factories, and their successes were based on carefully designed plans and strategies. The discoveries in science and technology have greatly led to tremendous success in the manipulation of material resources and human environments in favor of humanity. The resultant effect of scientific and technological innovations in medicine, engineering, meteorology, agriculture, management, economics, law, marine, aeronautics, in the three centuries have impacted tremendously on every aspect of human endeavor.

A World Bank report indicated that the first industrial

revolution in the United Kingdom took 58 years i.e. 1780 – 1838, while the adoption of improved technology took other countries less to attain the same feat. For instance in USA, it took just 46 years i.e. 1839 – 1885, Japan 34 years, South Korea 11 years, China 10 years, etc. This means that if we are committed to national development and nation building, we may still breakthrough technologically by 65th year of our independence.

More than ever in the past, the necessity for Africa to industrialize is being stressed at various international forums, ranging from TICAD VI to the G20 Summit, which put industrialization in Africa and Least Developed Countries (LDCs) in its programme for the first time. The 2030 Agenda for Sustainable Development and its Sustainable Development Goals (SDGs), in particular Goal 9, the Addis Ababa Action Agenda and the Third Industrial Development Decade for Africa (IDDA III) resolution, also mark a transition to a new development paradigm with the recognition that Africa has to restructure and diversify its economies to be on a sustained growth path. Africa can more than double its overall GDP per capita by increasing its industrial GDP in the next ten years. According to the theory of endogenous growth, one of the key factors generating fast growth is human capital accumulation through learning by doing, or on-the-job training, as well as education. Grossman and Helpman (1989, 1990) identified knowledge spillovers from advanced to developing countries as the most important gains from trade. In order to maintain a high rate of learning, people need to change jobs, which necessitates the continuous introduction of new industries and new products, requiring different skills and technologies. Government assistance can also affect the speed of human capital accumulation. The government's assistance can encourage private industrialists to undertake new projects by reducing the risk they face. Thus, the diversification of the industrial structure with governmental assistance enables the learning process to continue without being subject to diminishing returns. Korea has maintained one of the world's most competitive educational systems, in which access to higher education is determined by a uniform standard.

## Nigeria's Mini Industrial sector at present

Data from World Bank Investment Climate Survey (WB 2006) and Nigeria's National Bureau of Statistics (NBS various years) survey which covered manufacturing firms, micro-enterprises, retail, and residual businesses addressed a wide range of issues pertinent to the industrial sector in Nigeria. According to the survey, among the 2,387 firms surveyed, only 42% fell within the industrial sector. The distribution of firms across, age, ownership, and export status in the Nigerian manufacturing sector is dominated by firms in the food (30.17%) and garment (22.28%) sub-sectors—firms in the dominant sub-sectors over twenty years old have a relatively smaller percentage.

The constraints to growth as reported by the firms in the sample includes Electricity outages, transport bottlenecks, crime, and corruption are all key factors. Nigerian manufacturers suffer acute shortages of infrastructure such as good roads, piped water, and, in particular, power supply. Electricity outages and voltage fluctuations are commonplace, causing damage to machinery and equipment. Consequently, most firms rely on self-supply of electricity by using generators, escalating costs of production and eroding competitiveness relative to foreign firms. Only 3.7% of firms surveyed have access to formal credit and a phenomenal 38.9% pay bribes. Corruption, rent-seeking, and patron–client relationships impinge on the cost of doing business and contribute to a poor investment climate in Nigeria. Many firms are forced to offer gratifications to public officials for sundry purposes such as accessing public utilities, clearing goods at the ports, and obtaining licenses and permits. Credit delivery from the financial system circumscribes smaller and medium-sized firms. To bridge the gap, the government provides subsidized credit to favored sectors and firms. Implementation of these initiatives is typically faulty, with funds failing to get to the intended beneficiaries.

A prominent feature of the mini industrial sector in Nigeria is the existence of a number of special economic zones. There are approximately twenty-five free trade zones (FTZ) licensed by the federal government. However, fewer than thirteen of these are currently operational. Two types of free trade arrangement operate in Nigeria—specialized and general-purpose. These are managed by two bodies—the Oil & Gas Free Zone Authority for the oil and gas zone and the Nigerian Export Processing Zone Authority (NEPZA) for the general-purpose zones.

A self-reliant economy is self-sustaining and self-propelling. This requires that an industry in Nigeria is fed by agricultural raw materials preferably from local farms and plantations, or by minerals from mines located 'within the country; the machinery and equipment as well as spare parts are manufactured locally; and the universities are adequately staffed and equipped to produce the high level manpower to man the industries and research institutes.

In looking inwards, Nigeria is rich in scientific and technological manpower who have appropriately organized themselves into learned societies like the Nigerian Society of Engineers, the Chemical Society of Nigeria, etc. Nigerian scientists and technologists can build iron and steel industries, oil refineries, food processing industries, etc for the country if the Government of the country will mobilize, empower and commission them. Nigerian governments must look inward and use what they have to develop the country instead of looking outward to elusive and illusive foreign investors, for only Nigerians can develop Nigeria, foreigners will not. Because the indigenous private business sector is too weak and wobbly to initiate and carry through the onerous task of industrializing the country, it is imperative, at this level of the country's economic development, that Nigerian government build industries directly or in partnership with indigenous entrepreneurs in order to provide gainful employment for Nigerian citizens and create wealth to tackle the problems of broken down infrastructures, mass poverty and social services.

Government's capability in building industries is assured by its constitutional control of the national revenue and therefore can afford the necessary capital to finance scientific research and development and industry building. Research and development is sine qua non to industrialization and is the only pathway to Nigeria developing home-grown technology to overcome technological backwardness.

To ensure speedy utilization of R&D findings Government should produce on a regular basis Development plans in which it defines its industrial targets, articulates plans and strategies for attaining those industrial objectives, and articulates its overall economic development heights and infrastructural goals. R&D work without properly articulated national economic goals to which the R&D work is tied becomes a blind alley and an undue drain on the nation's purse.

If they abandon the industrialization of the country to the whims and caprices of "foreign friends," the country will never develop technologically and will never achieve economic independence. Lucas (1993) claimed that the quicker the introduction of new products into the economy, the quicker the process of learning by doing. The Korean experience strongly suggests that a government can play an important role in sustaining human capital accumulation through learning on the job and economic growth. Labor efficiency also depends on the educational curriculum, on-the-job and off-the-job training, labor-management relations, and work ethics. It is to be noted that the Asian NIEs placed high priority on vocational, technical, science-based education and training to produce an efficient industrial work force.

Nigeria operates on the principle that the country needs small and limited government. This doctrine stipulates that it is not the duty of government to build industries; it is the duty of the private business sector. Government only creates the enabling environment for private sector investment. In the Nigerian situation, this argument is untenable.

The indigenous component of the Nigerian private business sector is weak and wobbly as it suffers from severe constraints that stunt its growth and retard its progress. These traits include inadequate finance capital, lack of technological know-how, mutual distrust that inhibits the entrepreneurs from pulling their resources together to form large and more profitable concerns, and high cost of production arising from resort to the use of electric generators as a result of the grossly inadequate and epileptic power supply from the national grid. Obviously therefore the indigenous private business sector cannot industrialize Nigeria.

The foreign component is the foreign investors that have been vigorously wooed by successive Nigerian governments from independence in 1960 to date and have not come to industrialize the country. It becomes imperative that for Nigeria to industrialize, the governments, Federal and State, must take the lead by financing scientific research and development, and building and running industries directly or in partnership with the indigenous private sector. The logic of this case derives from the fact that

(i) The Nigerian government collect and disburse the national revenue from oil and other sources that run into trillions of Naira per annum and therefore can most conveniently afford the finance capital required for industrialization.

(ii) As a developing country it is foolhardy to imitate the advanced countries with a highly developed indigenous private business sector and advanced technology who, therefore, can afford the luxury of small and limited government.

The neocolonial status can be overcome. China and India, both former colonies, have escaped the raw materials supplier and foreign goods consumer status originally meant for them, and emerged industrial powers. Nigeria can also pull herself out of this syndrome. This she will achieve by looking inward and going massively into industrial manufacturing through home-grown technology. Only the governments can lead this crusade. Policy makers at all levels in Nigeria need to be keenly aware that few countries can achieve development goals of economic diversification, food security, improving health systems, cleaner energy, generating wealth and jobs, and reducing absolute poverty, without the scientific, engineering, and technical capacity to handle these challenges. There are no sustainable solutions if countries do not build the capacity to find and develop appropriate technologies, and modify them for local use. There is a need for an overarching national strategy for Science, technology and innovation: a strategy that will restructure our Science and technology sector for greater coordination, communication, and policy harmonization.

# CHAPTER 4:

## PANACEA TO OVERCOME NIGERIA'S PERSISTENT SLUMBERING

## 4.1. National Economic Development Blueprint (Blueprint Strategies)

Nigeria's prevailing industrial and technological backwardness derives directly and squarely from the fact that successive Nigerian governments from independence in 1960 to date have followed the colonial policy that relegates the country to producer of raw materials and consumer of foreign industrial goods. The governments' reason that the industrialization of the country can best come from foreign investors who possess the requisite capital and technical know-how. This reasoning, erroneous as it is, has matured into belief, and belief into policy. The Nigerian administration has a tendency to delay serious proactive actions in policy formulation until its hand is forced. In fact, rather than a conscientious blueprint for progress, most industrial development plans are no more than countermeasures to correct perceived deficiencies in the economic sphere. Invariably, such knee-jerk policies are severely limited in scope and cannot sustain a framework for serious long-term industrial development.

The Nigerian government must transcend the grandiose design of industrial blueprints that remain academic in nature and, therefore, impractical and unsustainable. It is of urgent national importance that Nigeria recognizes the emerging paradigm shift in technology in order to capture the benefits of these advancements. In the last decade, however, the productive sector has oscillated in the abysmal low single digits in contributions to the nation's gross domestic product, aided by a derelict infrastructure and perverse econo-political variables. The fact is that nations without a proactive industrial policy framework flexible enough to recognize and seize emerging technologies can never become relevant players in the industrial development march since they lose crucial comparative advantage by utilizing outdated technologies in production. To be sure, government always has a central role to play in the industrialization process, but it has become devastatingly obvious that the Nigerian government has been involved far beyond the call of duty. The incestuous tripartite relationship between government, business moguls, and ethnic considerations in formulating the indigenization policy has been very troubling. The result of this intercourse of economically adverse players has been bureaucratic corruption that has eventually matured into a systemic monster that refuses to be tamed. The Nigerian administration must institute a divorce in the perverted marriage between government, business moguls, and ethnic considerations. Unambiguous policy formulation must be devoid of these unhealthy developmental distortions. A translucent and workable industrial policy may still remain inadequate, however, due to deficient human capital, funding, and base industrial complexes. While the importance of an enabling environment can never be stressed enough, no tangible developmental gain can be sustained if other government policies are adverse to industrial growth. As noted earlier, it's imperative that any techno-industrial policy formulated by the government is robust enough to identify new and emerging technologies at their nascent stages to allow integral engagement in the development process.

## Historical Review of Past National Economic Development Plans in Nigeria

At independence in 1960, and for much of that decade, agriculture was the mainstay of the Nigerian economy. The sector provided food and employment for the populace, raw materials for the nascent industrial sector, and generated the bulk of government revenue and foreign exchange earnings. Following the discovery of oil and its exploration and exportation in commercial quantities, the fortunes of agriculture gradually diminish. At independence, the contribution from the primary sector to GDP was about 70 per cent. The transition from primary production to secondary and tertiary activities was sluggish; in 2009 more than half of Nigeria's output was still generated by the primary sector. The secondary sector contributes the least to GDP in Nigeria.

Many situations in Nigeria defy conventional models and ignore globally accepted norms to such a degree that they have become a travesty. In fact, numerous well-intentioned policies often end up implemented in such a way that they defy their original purpose. For example, law enforcement equipped with taxpayers' funds often malfunctions and becomes a threat to the very citizenry it was empowered to protect.

At Post-independence, Nigeria's first attempt at comprehensive and integrated planning took the form of the First National Development Plan (1962–1968). The plan included an aggregate growth rate target of 4 per cent per annum, an increase in the rate of investment from 11 per cent to 15 per cent of GDP and an increase in the 'directly productive component' of government investment (Bevan et al. 1999: 30). To encourage industrial development and lessen dependence on foreign trade, import substitution industrialization (ISI) was introduced conserving foreign exchange by producing local products that were previously imported. Import duty relief, accelerated depreciation allowances, and easy remission of profits aimed to attract foreign investors. The period of this plan witnessed the commissioning of energy projects such as the Kanji Dam and the Ughelli Thermal Plants, which provided a vital infrastructural backbone for the emerging industrial sector. Other important industrial infrastructure included an oil refinery, a development bank, and a mint and security company. Clearly, the main objective of the ISI strategy was to stimulate the start-up and growth of industries, as well as enhance indigenous participation. However, it led to high technological dependence on foreign know-how, to the extent that the domestic factor endowments of the country were grossly neglected.

Post-civil War Oil Economy era saw the Second National Development Plan (1970–1974) emerge. The government embraced ambitious and costly industrial projects in sectors such as iron and steel, cement, salt, and paper (Oyelaran-Oyeyinka, 1997). The period of the 1970–1974 plans also witnessed a dramatic shift in policy from private to public sector-led industrialization. It was clear that there was a dearth of human capital and skills required for initiating, implementing, and managing industrial projects among Nigerian entrepreneurs. Foreign technical skills and services were heavily relied upon. The oil economy was characterized by 'Dutch Disease', signified by the diversion of productive resources away from agriculture into commercial activities that thrived on trade in imported manufacturing goods (Forrest 1993). The windfall in oil revenue affected the fiscal policy of government. Political pressures meant that the tax base remained narrow under the belief that oil revenue would always lead to surplus. Non-oil taxes were thus neglected and some taxes abolished.

The Third National Development Plan (1975–1980) followed suit, this was launched at the height of the oil boom—emphasis remained on public sector investment in industry.

Indigenization policy was implemented in 1973 and 1978, with the objectives of increasing the level of local managerial control, building local technological capability, and extending state ownership. Heavy subsidies were provided for public companies and corporations. The Nigerian Enterprises Promotion Act of 1977 aimed to further support Nigerian businesses. It became apparent that the country had entered into industrial project agreements with very little concern for capabilities for technology acquisition. While each of these projects required the acquisition of key sector-specific skills, the agreements made by Nigerian planners were for the turnkey transplantation of technology. During the same period, the nation's oil sector had become vibrant and prosperous, and the gates of the economy had been opened up to all sorts of imports. This had a debilitating effect on real industrial growth. The period of the Third National Development Plan failed to advance the course of industrial development in Nigeria in a positive way.

The Fourth National Development Plan (1981–1985) coincided with a global economic recession which sparked declining foreign exchange earnings, balance of payment disequilibrium, unemployment, and accelerating inflation in the Nigerian economy. This prompted emergency stabilization measures in 1982. These measures included advance deposits for imports; increases in import duties; review of import licenses; a 40 per cent across the board cut in public expenditure without any prioritization; and an upward review of excise duties, interest rates, and prices of petroleum products. In the agricultural sector, exports became highly constrained by the overvalued naira and production declined. This included the production of labor-intensive export crops (e.g. cocoa, palm oil, cotton). The decline in output was most apparent in the manufacturing sector, resulting in gross losses in employment. This demonstrated the vulnerability of the high cost, import-dependent industrialization that had been encouraged by the pattern of incentives in the 1970s. A decline in the aggregate index of manufacturing was observed from 1982, falling by 26 per cent in 1983 (Forrest 1993). Plant closures were common in consumer goods sectors, especially in textiles. Average capacity utilization in industry declined from 73.3 per cent in 1981 to 38.2 per cent in 1986 (Fashoyin et al. 1994). The stabilization measures achieved some reduction in the volume of imports, however, the inability to effectively control the allocation of import licenses and foreign exchange aggravated the pace of decline. The experience in the first half of the 1980s exposed profound weaknesses in Nigeria's industrial structure and planning.

For the Fifth National Development Plan(1999-2007), the role of Science and technology featured prominently in the economic reform agenda, specifically within the rubric of Vision 20:2020, the National Economic Empowerment and

Development Strategy (NEEDS 2004-2007); 2017 National Economic Recovery and Growth Plan (NERGP); Roadmap for Growth and Development of the Nigerian Mining Industry (2016); the Nigerian Industrial Revolution Plan (2014); the Agriculture Promotion Policy (2016-2020); the National Renewable Energy and Energy Efficiency Policy (NREEP, 2015); the National Health Policy (2016); the National Communication Technology Policy (2012); the Draft National Transport Policy (2010); the Nigerian Water Sector Roadmap (2011); and the Roadmap for the Nigerian Education Sector (2009). One of such short-term plans is the National Economic Recovery and Growth Plan (NERGP, 2017-2020) that focuses on the following objectives: macroeconomic policy improvement, economic diversification, competitiveness improvement, social inclusion, and Jobs creation but fail to make any positive gains to economic development of the country.

Similarly, the Sixth National Development Plan (2008-2015) current economic policy blueprint—Nigeria Vision 2020 (NV20:2020) (NPC 2009)—embraces elements of science, technology, and innovation (STI). With all the aforementioned Nigeria's national economic agendas since independence in 1960, science and technology was never given a central role, neither was it considered a driving momentum for economic transformation and development where all the other sectors of the economy will branch from. It was mostly considered as a sub-sector or an element of the whole economy. That is why all the plans have failed even before they started.

### 4.1.1. Overview of NIPF-Nv20:2020 and NSTIR 2017-2030

### (A general Overview of the Nigeria's Industrial Policy Framework-Nv20:2020 and Nigerian Science, Technology & innovation roadmap 2017-2030)

The persistent failure of Nigeria's industrial roadmaps may be indicative of an inherent defect in the policy framework or, at the very least, in the manner of implementation. Keen observers of Nigeria's industrial journey agree that poor implementation of government-declared policies remain an inexcusable factor in the failure of intended industrial development objectives.

### (1)Nigeria's Industrial Policy Framework- Nv20:2020

The Nigeria's industrial policy framework- nv20:2020 industrialization strategy aims at achieving greater global competitiveness in the production of manufactured goods by linking industrial activity with primary sector activity, domestic and foreign trade, and service activity. A key component is the promotion of a comprehensive policy of cluster development in the manufacturing and processing industries. This includes the development of industrial parks, industrial clusters and enterprise zones, and incubator facilities.

Industrial parks, aimed at large manufacturers, are expected to cover areas of more than 3,050 km2. The parks will be based on the comparative and competitive advantage of each geographical zone. The following business activities have been identified for each of the zones (Nigeria Vision 20:2020).

North East: agriculture and solid minerals e.g. gypsum, biomass, ethanol, biodiesel, tropical fruits, etc.;

North West: gum Arabic, livestock and meat processing, tanneries, biofuel, etc.;

North Central: fruit processing, cotton, quarries, furniture and minerals, boards, plastic processing, leather goods, garments, etc.;

South East: palm oil-refining and palm tree-processing into biomass particle boards, plastic processing, leather goods, and garments;

South West: manufacturing (especially garments, methanol, etc.), distributive trade, general goods, plastic, etc.;

South central: petrochemicals, manufacturing (plastic, fertilizer, and fabrications, etc.), oil services, and distributive trade (TINAPA).

The industrial clusters, which will be established with the participation and assistance of states and local governments, will cover areas of between 100 and 1,000 ha. They will be exclusively devoted to the organized private sector. The location of the clusters will take into account access to roads, railways, sea ports, cargo airports, and proximity to a city and management will be through a private cluster company. Industrial incentives similar to those in industrial parks will also be provided, while each cluster will have a skill acquisition/training centre.

Enterprise zones are platforms of 5–30 ha, targeted at incorporating the informal sector into the organized private sector. Located in state capitals and local government areas, they will enable farmers and SMEs to feed their products into the value chain of large-scale industries. These centers will accommodate mechanics, block makers, small-scale furniture manufacturers, timber merchants, and other vocational workers who constitute over 70 per cent of Nigeria's private sector. Skills acquisition/training centers will also be located in each enterprise zone, while management will be handled by the private sector.

The incubators will be start-up centers for new and inexperienced entrepreneurs, graduates of tertiary institutions, investors, and vocational workers wishing to set up their own businesses. In these centers, prospective start-up companies will be equipped with entrepreneurial skills and resources aimed at nurturing them from formation to maturity.

Nigeria still uses import prohibition to protect its manufacturing and agricultural sectors. The rationale is that the production base is relatively weak, import-dependent, and limited in technological capability. The import prohibition list includes a wide range of manufactured consumer goods that were often dumped in Nigeria's relatively large market. A few agricultural products (e.g. fresh fruits, pork, and frozen poultry) that are produced locally in large quantities are also included in the list to protect the local industry and encourage job creation. On the export prohibition list are staple foods/crops that are important for food security, commodities that could serve as raw materials to local industries and living organisms that are becoming rare. Such commodities include maize, hides and skin, scrap metals, and wildlife animals classified as endangered species

## (2) Nigerian Science, Technology & innovation roadmap 2017-2030

The primary objectives of NSTIR 2030 are said to include the following

1. to provide a long-term science and technology framework and support mechanisms for industrial revolution in Nigeria;

2. to facilitate the creation and acquisition of knowledge for production, adaptation, replication, and utilization of technologies to support Nigeria's technological and sustainable development aspirations;

3. to support the establishment and strengthening of organizations, institutions, structures and processes for rationalization of decision making; coordination and management of STI activities within an institutionalized national innovation system; and

4. to encourage and promote the creation of innovative enterprises that can beneficially utilize Nigeria's indigenous knowledge and technologies to produce marketable goods and services that compete with others in the global market.

5. Additional objectives of NSTIR 2030 are to coordinate and support the development of science and technology infrastructure to enable significant research for production of methodologies, models and data to support Nigeria's socio-economic development plans;

6. to devise and implement systems for identification and pruning of STI talent at all ages and educational levels in Nigeria through support and incentives to build a strong long-term workforce;

7. to coordinate the planning and catalyze the implementation of strategic projects such as those of space exploration, advanced computing, telemedicine, robotics, advanced navigation systems and, nanomaterials that can accelerate the emergence of Nigeria as a technologically developed country.

Some salient points of the roadmap worth noting are observed as follows

(A) That Nigerian pupils don't usually want to study Mathematics; it said the situation had been a setback for infrastructural development and industrialization.

(B) The above development had always compelled the country to contract major projects such as the building of refineries, roads and bridges to foreign companies, pointing out that the development has culminated in huge exportation of jobs from Nigeria.

(C) The roadmap had become necessary in view of attendant unemployment rate, worsening poverty and inability to build a self-reliant nation. It is meant to "Take our people beyond where we are."

(D) The blueprint had become imperative in view of the recent projection that in 2050, Nigeria would be the third most populous country in the world and the largest economy in Africa, affirming that without building a commensurate technologically inclined country, the projection would remain an illusion.

(E) In essence, the difference between the developed nations of the world that are the richest and developing countries that are poor is science and technology.

(F) Roadmap would also provide the platform to educate Nigerians that the country had a very rich science and technology heritage, recalling that the ancestors created artifacts, fine works of art.

(G) That the ministry would harness invention and innovation in order to catapult the nation towards economic recovery.

(H) That local contents would now be encouraged in all Science and Engineering contracts in the country.

(I) there was an appeal to ministry of Science and Technology staff to study and be guided by the Roadmap in their Scientific and Research endeavors.

(J) Unlike past policy guidelines, the Roadmap was infused with practicable timelines to boost Scientific, and Technological Development of the country.

(K) The need to move our economy from a resource-based to knowledge-based and innovation-driven economy cannot be over emphasized at this point in time. It is an undeniable fact that any nation that aspires to greatness cannot but make science, technology and innovation (STI) the linchpin of its aspiration.

(L) That the Nigerian industrial sector contributes only about 3 per cent of Nigeria's export revenue but gulps over 59 per cent of Nigeria's import, thereby negatively affecting the country's balance of payments.

(M) That the above mentioned has resulted in high levels of unemployment, poverty, hunger, poor healthcare service and high rate of illiteracy among others. This situation cannot be allowed to continue. The national science, Technology and Innovation Roadmap (NSTIR) 2030 was conceived as Nigeria's strategic plan to catalyze Nigeria's long-term sustainable economic development in consonance with the national policy on science, technology and innovation.

(N) That the road map intends to utilize STI to drive economic growth and diversify the economy. It will also promote commercialization of our R&D results, create employment and wealth, reduce poverty and support the realization of the goals of the Nigerian Economic recovery and Growth Plan (NERGP)

2017 – 2020.

The NSTIR 2030 with respect to implementation is divided into 7 categories of objectives, each of which comprises several initiatives and projects. The 7 categories which align with the Roadmap's objectives are Science Policy Support Programmes and Activities;

Science and Technology Improvement;

Research and Development Intensification;

Training and Talent Deployment;

Technology Deployment and Commercialization; and

Science Literacy Improvement and

Public /Stakeholders Engagement.

The NSTIR 2030 will be implemented in three time segments, namely:

(A) Short Term (2017-2020);

(B) Medium Term (2021-2025), and

(C) Long Term (2026-2030). NSTIR 2030 covers many high-utility projects that will be implemented by the various institutes/centers of FMST in collaboration with industrial partners, universities, other government entities and NGOs. Examples are

(i) commercialization of locally invented equipment and products, (ii) establishment of the National Science and Technology Agency/Fund,

(iii) implementation of artisan training programmes,

(iv) manufacturing of another set of satellites with expanded involvement of Nigerian scientists and engineers, establishment of advanced analytical laboratories and fabrication of several equipment and their components.

(v) That unlike past policy guidelines, the roadmap was infused with practicable timelines to boost scientific, and technological development of the country.

The aim of the formulation of this important document was to encourage our youths develop interest in mathematics and science subjects as well as reduce our level of over dependence on imported and foreign goods and services in Nigeria. the Road Map targets the mobilization of Nigeria's intellectual resources for the growth and diversification of the economy, provision of incentives for all stakeholders, to embrace and engage in science and technology innovation in order to improve science structures intensify and develop skills, deploy and commercialize technologies and improve science literacy engagement processes in Nigeria.

(vi) Research and development support will be given by FMST units to steel development, automobile production, implementation of renewable energy technologies, telemedicine, local drug manufacture, processing of agricultural products, development and application of new materials in infrastructure and individuals processes, and development and economy-wide applications of ICT techniques, as well as several other STI advancements.

As described in Nigeria's Industrial Revolution Plan published in January, 2014, systems are planned to make industry the dominant job creator and income generator up to 2020. The plan which is outlined covers the creation of 8 general-purpose specialized industrial cities in strategic locations along transport corridors, creation of 6 Technology Innovation Clusters and improvement of services at Nigeria's 27 Free Trade Zones. These facilities will present more opportunities for science and technology-catalyzed industrialization and create jobs.

Budget estimates for the short term programme total N180 billion over the three budget years (4-year duration) with the distribution of Programme Configuration and Planning (1.5%), Stakeholder Engagement Processes (2.7%), Management and Personnel Support (11.6%), Facilities and Equipment (25.6%), Deployment and Diffusion of Deliverables (3.4%) and Project Operations (55.2%). NSTIR 2030 will be implemented in collaboration with a wide variety of stakeholders, including academic institutions, public and private research and development centers, the private sector, State and local government agencies, non-profit and community groups, development partners and professional associations using revised and more efficient structures and governance systems that have been ratified by the Federal Government of Nigeria through the Federal Ministry of Science and Technology.

# Observations and Shortcomings of the Industrial policy frameworkNV20:2020 and NSTIR 2017-2030

1. The Nigerian science, technology and innovation roadmap 2017-2030 policy should have been brought forward first before the Nigerian industrial policy framework of NV20:2020. A child after being given birth to, does not start with running steps before going back to crawling, rather he learns how to crawl first before walking and finally running. This is because there must be an initial grand deliberate attempt at sensitization, educational restructuring, training and mobilization of requisite skilled manpower as postulated in the NSTIR 2017-2030 which will be the driving force for scientific and technological revolution before the next phase which is introducing the industrial policy framework NV20:2020 that will lead to the industrialization of the country.

2. The Promotion of a comprehensive policy of cluster development in the manufacturing and processing industries as outlined in NV20:2020 Industrial policy framework will not open the entire country to massive infrastructural development that Nigeria so much needed as witnessed by the Asian tigers wonders. Some regions will be at disadvantage than others and will therefore not benefit from equal income distribution package which is a prerequisite for rapid industrialization.

3. It is not just enough to have science, technology and innovation knowledge you need to outline the areas of application of the knowledge gotten or the innovations, that is

in industrialization through production and manufacturing.

4. Both the Industrial policy framework Nv20:2020 and NSTIR - 2030 are not explicit about the type of policy approach it intends to carry, is it an export-oriented policy? Which encourages exportation of industrial goods produced from the result of knowledge gained from STI or is it a policy of local produced-local consumed or strictly imports substitution policies?

5. What areas of competitive advantage does the policies intend to specialize on? Both policies failed to identify, recognize and specify those important areas of specific advantage or areas of concentration for the projected finished products of industrialization. For example, Switzerland specialized in Banking institutions and watch making, Japan specializes in electronics and automobiles with global brands like Sony, Yamaha, Toshiba, Toyota, Honda, Suzuki etc, the U.S.A specializes in automobiles like GM motors, Ford, D. Chrysler, Hummer, and military hardware like Abraham Tanks, B2 Stealth Bomber, B52 Bomber, F16s, Apache Helicopters, Tomahawk Long range guided missiles and South Korea specializes in Home consumables like LG, Samsung etc.

6. The NSTIR 2017-2030 failed to address the very important issue of income distribution from the success of industrialization; one of the main success drivers of the Asian Tiger economy is their seriousness and ability to distribute income equally throughout the country.

7. All the rhetoric in the NSTIR 2017-2030 and previous Roadmaps which had failed to make any gain or impact to the Nigerian economic and socio-political landscape because it is termed "Roadmap" and planned as such was just a policy guideline and is indicative of none proper planning. It should have been termed "strategies" equipped with full chronological events and targets which are attached to feasible future goals to be achieved.

8. The NSTIR 2017-2030 should be seen as a national economic blue- print and not just a sectoral or ministerial target. Science, technology, innovation should be an important integral national economic development plan and not be separated from the economy. Example, the national science, technology and innovation roadmap NSTIR 2017-2030 is controlled by or under the Federal ministry for science and technology while the Roadmap for Growth and Development of the Nigerian Mining Industry -2016 is controlled by or under the Federal ministry for Mines and solid minerals development.

9. The Nigerian Industrial revolution plan published in January 2014 for vision 2020 is not necessary; too much publication of what to do and not doing it is a waste of time. Since 2014 to now 2018, no single general-purpose specialized industrial cities were created in any strategic location in Nigeria neither was there any single technology innovation clusters created as stated and mandated by the document. The policies lacked procedural chronological strategies. With respect to history, the

industrial revolution that took place in Britain in the mid 18$^{th}$ century and latter in America by the end of 19$^{th}$ century was not pre-written like we have it now in Nigeria, it took place automatically because the necessary atmosphere prevails and the players involved acted appropriately when the opportunity presents itself. The inventors and industrialist of the early period did not wait for government of the day to write to it any roadmap guidelines for invention, innovation or industrialization.

10. Both policies guidelines failed to recognize, prioritize and allocate a specialized rapid technology transfer center where advanced foreign products of all kinds are imitated or modified by opening, dismantling and mapping out holistically by indigenous scientists and technologists down to the building block architecture to identify, familiarize, modify and advance the products for better local consumption, adaptation and durability. This center will generate fast insight into blueprints for PCBs, or CB, AI( if any used), type of programming technology and software used, energy type and source, type of mechanism and mechatronics, hardware material and its chemical composition, type of fabrication and welding/joinery, cooling and heating systems etc. This was now general practice in advancing technology countries like Israel, China, Brazil, India, Mexico, Iran and so on.

## Overview of American Marshall Plan

The Marshall plan have come to become a household name globally in the late 40s and 50s after the Second World War which saw the axis powers' total defeat and unconditional surrender. In the general sense, the term referred to a grand strategy for assistance and recovery. Otherwise known as European Recovery Program (ERP), it is the United States programme of financial assistance that helped to rebuild European countries devastated by World War II. It was commonly known as the Marshall Plan, named after US Secretary of State George Catlett Marshall. After the war, Europe's agricultural and coal production had nearly stopped, and much of the population was threatened with starvation. The Europeans also lacked dollars which would enable them to purchase new materials and machine tools from the United States to help restore their shattered economies. The United States responded for four reasons. First, Europe had been a large market for American goods, and without a prosperous Europe, the United States might have suffered a severe economic depression. Second, without American aid, Western Europe might succumb to communism, and US leaders considered that prospect a threat to American security. Third, Western Europe appeared open to influence by the Union of Soviet Socialist Republics (USSR), which the United States was beginning to see as its main rival. Fourth, West Germany (now part of the united Federal Republic of Germany), historically the continent's industrial hub, had to be rebuilt as a buffer against Soviet expansion; European fears of their World War II enemy would lessen only if the Germans were integrated into a larger Europe. After careful planning, in June 1947 Marshall announced that if Europe devised a cooperative, long-term rebuilding programme, the United States would provide the necessary dollar funds. Great Britain and France called other Europeans, including the Soviets, together at Paris. When the Soviet delegates learned that the United States insisted on close Soviet cooperation with the capitalist societies of Western Europe and an open accounting of how funds were used, they left Paris and established their own plan to integrate Communist states in Eastern Europe. An economic curtain divided the continent. The Congress of the United States distributed more than $13 billion in aid. Seventy per cent was spent on goods in the United States. The Economic Cooperation Administration distributed the money, and the Organization for European Economic Cooperation (OEEC) spent it. The largest amounts went to Great Britain, France, Italy, and West Germany, in that order. As Cold War tensions rose in 1949, the funds increasingly went into military expenditure rather than industrial rebuilding.

The programme achieved both its immediate and long-term aims: When the aid ended in 1952, Communist control of Western Europe had been averted, the region's industrial production stood 35 per cent above pre-war levels, and West

Germany had become independent, and her economy was beginning to recover rapidly.

In like manner, Nigeria should borrow a leaf from the strategy that rebirth a devastated war torn Europe and reinvigorated its economy into prosperity through the implementation and sustenance of a Marshall Plan style blue print to be built on a sustainable platform of science, technology and innovation which will catapult the nation to industrialization and export driven economy.

### 4.1.2 Proposed NRMP 2019-2040- Blueprint

### (Proposed Nigerian Revolutionary Marshall Plan (NRMP) 2019-2040 -The Blueprint)

1. Science, Technology, and Innovations sustainable strategies (STISS) – 2019-2026

2. Infrastructure Development Investment sustainable strategies (IDISS) – 2019-2025

3. Industrialization sustainable strategies (ISS) – 2026-2033

    (A) Technical Aspect:

    (B) Industry Trade Aspect:

4. Service Based Economic sustainable strategies (SBESS) 2037-2040

# Breakdowns of the Proposed Nigerian Revolutionary Marshall Plan - Blueprint

## 1. Science, Technology, and Innovations sustainable strategies (STISS) – 2019-2026

A. creation of a national academy of science (NAS) – to ;
(i) Considering the high priority of this body to national development, the national academy of science shall by law establishing it be a direct arm under the presidency and the chair of the national academy of science shall be empowered to directly report and advice the president of Nigeria.
(ii) The national academy of science shall be composing of seven (7) seating bench members who shall be sorely responsible for electing their chair.
(iii) maintain consistent updated science journal,
(iv) organize scientific bench hearing on postulated theorem(s), breakthrough invention, innovations, science papers and patent claim,
(v) organize independent scientific verification and experimentation on submitted postulated theorem(s) and patent claims,
(vi) publish tested theorems approved into law by the apex body

after experimentation and verification,

(vii) approve and adopt experimented and proven patent claim right and sent same to the executive for ratification through the legislature,

(viii) organize or approve some specific research teams and recommend them for research funding and grants,

(ix) initiate, formulate and approve school science and technology curriculum at all levels,

(x) maintain a register of genuine national scientists, technologists, entrepreneurs, innovators and so on in both the public and private sectors.

(xi) Responsible for organizing both national and international science and technology expositions in collaboration with organized private sector, key industry players and international partners,

(xii) approve scientist, technologists and innovators for science and technology exchanges, technology transfers and overseas training between and amongst countries,

(xiii) organize science competition amongst schools in various categories for primary schools pupils, junior and senior secondary schools level and for tertiary institutions level.

(xiv) Responsible for nominating individual works, groups and Team collaborative projects and institutions projects for local and international awards and rewards.

B. Identify specific areas of competitive advantage vis- a -vis locally available and abundantly sourced raw materials, cultural consideration, superior technical skill inclination, area of national interest leading to location of science facilities.

C. Construction of science and technology Research & Development Centers ;-

(i) Two(2)large state- of –the- art science laboratories, completely furnished with all modern science equipments in each of the six(6) geo-political zones of Nigeria,

(ii) One(1) medium size state -of- the -art science laboratory in each of the 37 states of Nigeria and the FCT, completely furnished with all modern science equipments.

(iii) Construction of six(6) new "specialized rapid technology transfer center", with all supporting advanced testing and modification equipments, one (1) in each of the six geo-political zones of Nigeria.

(iv) Two(2) large state-of-the-art Practical Technology development complexes (PTDC) complete with cutting-edge machines and equipments for practical application and tests in each of the six(6) geo-political zones of Nigeria strategically located in areas of advantage.

(v) One(1) small size state -of- the -art Practical Technology development complexes (PTDC) complete with cutting-edge machines and equipments for practical application and tests in each of the 37 states of Nigeria and the FCT, situated in areas of advantage to the projects.

(vi) Two(2) state- of -the- art Innovation showrooms (IS) ,furnished with all necessary and modern showcasing hardware and 3D presentation gadgets and facilities in each of the six(6) geo-political zones of Nigeria,

(vii) One (1) Hi-Tech Simulation center (HTSC) for each of the six (6) geo-political zones of Nigeria, completely furnished with up-to-date simulation apparatus for virtual real-time realities for both sea, air and land gears.

(viii) One(1) small technology and new innovation exposition centers(TNIEC) located in each of the six(6) geo-political zones of Nigeria, and

(ix) Two(2) large international technology and new innovation exposition center(ITNIEC) located one in Nigerian capital Abuja and one in Lagos.

D. Organize several and various unique and independent science and technology research teams or group's project teams with each having specific objectives and incorporating various disciples input to the project to achieve the desired goal. Examples;

(i) Team A-based in location Y due to the advantage of the location to the project is a science research team project with Joint government and company V funding to engage in a project to 'develop a nano printed circuit board(NPCB) to be used in Planes autopilots system has-

Mr. T as its project manager,

Mss. H as its team co-coordinator,

Mr. L as its computer programmer analyst,

Mrs. O as its genetic engineer,

Mr. X as its Material scientists,
Prof. R as its Lead Researcher,
Engr. J as its Electrical engineer,
Dr. C as its Structural engineer,
Prof. U as its Liaison anchor,
Dr. A as its Industrial Physicist.
Etc

(ii) Team B-based in location Y due to the advantage of the location to the project is a science research & Technical based project with government funding conducting research on 'first Nigeria indigenous lower earth orbiting artificial satellite' has-
Prof. M as its Project Manager,
Engr. T as its Aeronautics Engineer.
Mss. K as its team Director,
Mr. P as its computer designer & programmer,
Mrs. C as its Aerodynamics engineer,
Mr. Z as its Robotic engineer,
Prof. W as its computer designer,
Prof. N as its Mechatrionics expert,
Dr. B as its drone engineer,
Mr. X as its Lead Researcher,
Engr. V as its Communication specialist
Dr. Mrs. D as its Instrumentation engineer,
Prof. Q as its Electrical & electronics engineer,
Dr. C as its Structural engineer,
Mr. E as its Liaison anchor,

Prof. S as its Deep space engineer,

Engr. K as its Line Simulation engineer,

Engr. R as its Rocket Launch expert,

Engr. A as its Building engineer

Dr. L as its Applied Physicist

Engr. C as its Energy and Power engineer

Prof. U as its Industrial Chemists

Mrs. G as its Quantum Physicist

etc

E. First indigenous design blue prints (IDB) for production of; -
(i) 100% entirely new components (e.g. indigenous engine model, indigenous car model, indigenous smart phone model, indigenous sewage and water treatment plants, indigenous 3D Printer, Develop new computer language etc) using 100% locally sourced raw materials.

(ii) 60-85% innovation of new components from existing theorems (e.g. in ICT using the existing computer programming Languages like C++, Java, C#(C sharp)to design new set of useful system or applications software like operating system (OS) for smart phones, Tablets, PC, Orbiting Satellites, Laser guided missiles or executable applications software useful in specific disciplines like Genetic Mapping, Finance, Accounting, Structural engineering, Quantity surveying, Music, Computer Gaming etc)

(iii) 25-50% modification of existing components (e.g. exposing building blocks of existing components and consciously modifying most of the building architecture to create a different functional behavior)

these ready-made design blue prints, Streaming from all areas of scientific and technological research initiatives from individual science and technology researchers, team and groups projects and institution projects from all fields and disciplines must be classified and well guarded against intellectual property thieves, hackers, industrial espionage and piracy. The scientists, innovators, technologists etc themselves must have been initially compelled to swear an oath of secrecy and abide by it, while government on its part must provide adequate security to them and their immediate families.

## 2. Infrastructure Development Investment sustainable strategies (IDISS) – 2019-2025

### (a) National power investment revamping blueprint (NPIRB) – 2019-2024

This is an indispensable precursor to industrializing Nigeria. There should be a huge government investment spending in the power sector through new technology ventures in the sustainable renewable energy sources throughout the value-chain to boost the present national grid generation which stands at +_ 4000 Megawatts which is grossly inadequate for a large country like Nigeria and it scare away businesses and foreign direct investment.

(i) Existing Hydro and thermal power generation------------------------------- 4000 Mw

(ii) Proposed Solar energy investment (2019-2023) ------------------------- 2200 Mw

(iii) Proposed Wind Turbine generators energy investment (2019-2024) --------1700 Mw

(iv) Proposed Biomass energy investment (2019-2025) ------------------- 2000 Mw

(v) Proposed Geothermal energy investment (2019-2025) ---------------------- 600Mw

--------------

10, 500 Mw

_____

(vi) Proposed underground transmitting system (UTS); A total overhaul of the transmitting system in the national grid must be carried out to resemble that of developed economies. The underground transmitting system should be adopted just like the system in Abuja Municipal where no electric poles are seen around. This has far reaching advantages than the overhead surface transmission. The underground transmitting system should use armored cables and well computerized with digital real-time automated substations having mapped grid of every transmission line in every location in the country. This will prevent theft of transmission lines, power outages as a result of lightning and thunder, Wind negative action on electric poles and lines, illegal connections, lost of power through transmission etc.

## (b) National Continuous Light & Heavy Rail system Investment Blueprint (NCRSB) – 2018-2022.

Even now railroads carry far more freight in the United States than do trucks, barges, or any other form of transport; they are the backbone of passenger-travel systems in Europe; and they account for more of China's infrastructure investment than airports or roads. The renovation of the nation's railway system needs to be continued and indeed connecting all major cities by rail will in the end be a worthwhile investment. Rail travel is not only safe and convenient but also inexpensive. Re-introduction of an efficient rail service will not only promote interstate commerce, but will of necessity, decongest our poorly maintained roads that have become killing fields over the past 20 years. In this time of national reconstruction, we must borrow a leaf from the new deal era of Franklin D. Roosevelt of the USA and create an efficient network of roads to crisscross the entire country.

(i) Construct light gauge train service network linking all the 37 states of Nigeria including the FCT Abuja. The light gauge should transport mostly people to and from various locations in the country. This grid will use fast moving business passenger couches termed 'express trains' which do not stop at local stations as well as local trains which stops at every county or local station. The 'local goods train' should also run on these tracks to reach every center in the country where raw material is accessed. Raw materials transported through this local goods train should be taken to the storage facilities at the headquarters of the six geo-political zones where the heavy goods trains will transport it to the respective manufacturing industries.

(ii) The heavy rail grid should operate on standard gauge system; this should have more spans for wider couches to convey many raw materials from where they are sourced at the geo-political headquarters to the industries and from the industries to convey industrial finished products to the sea ports for export. The type of couches termed 'goods train' should be linked to all industrial complexes in the six (6) geo-political zones of Nigeria.

## 3. Industrialization sustainable strategies (ISS) – 2026-2033

## (A) Technical Aspect:

Designs of complete blueprints for modern industrial line production machines such as;-

(i) design of indigenous blast furnace and industrial hot moulds

(ii) design of indigenous industrial fabricators, Lathes design and cutting machines

(iii) design of indigenous industrial robotics and other automated production line machines

(iv) design of indigenous automated molding, cutting, fitting, shaping and sizing machines.

(v) design of indigenous mechatronics machines and 3D printers

(vi) design of indigenous industrial consoles containing digital and analogue meters for both quality and quantity assurances

(vii) design of indigenous industrial curing, drying, sterilization, spraying

(viii) design of indigenous industrial horizontal and vertical conveyor belt systems, lifts, cooling and heating electromechanical systems

(ix) design of indigenous industrial simulators

(x) design of indigenous industrial pressurize and momentum impact test machines and control environment.

(xi) Complete blueprints for design of indigenous Smartphone, engine system, car model, household items etc.

(b) Organized production, fabrication and creation of the indigenously designed industrial machines and equipments from the submitted blueprints.

(c) (i) construction of one(1) industrial manufacturing complex in each of the 37 states of Nigeria and the FCT, considering the advantage of proximity to available raw materials

(ii) construction of two(2) large warehousing complex in each of the six(6) geo-political zones of Nigeria within approach distance to the standard gauge railway grid.

(iii) Construction of one (1) medium size warehousing complex in each of the 37 states of Nigeria and the FCT, within approach distance to the single gauge railway grid.

(iv) Construction of one (1) national 'composites' manufacturing complex to international size and standard. This will gradually replace steel production from Ajaokuta steel rolling mill.

(v) Revamping of Ajaokuta steel rolling company for steel production

## (B) Industry Trade Aspect:

(i) Nigeria must destroy or abolish the traditional social hierarchy in the society; this will play a major role in motivating the citizens to invest in human capital. Such an environment of equal status with fair competition will create great potential for vertical mobility in society.

(ii) The government should also provide substantial tax incentives to exporting firms by reducing business and corporate taxes on export income by 50% and exempting tariffs on materials or intermediate goods imported as export content.

(iii) Nigerian banks should act as a treasury unit, the industrial sector should act as production and marketing units, and the government should act as a central planning and control unit.

(iv) Government's support to exporting firms should be based on export performance.

(v) To get more privileges, exporters had to work hard to compete with each other and foreign businesses. In this way, the Nigerian government can maintain an efficient allocating device for picking winners and will be able to reduce the risk of an "interventionist approach" Furthermore, this strategy will compel Nigerian firms to compete with foreign firms and bring tremendous externalities of accelerated learning on the job and, thus, a shortened learning curve.

(vi) Commercial loans should be mostly allocated to the manufacturing industries, while public loans should be allocated mainly to the service industry (mostly for infrastructure).

(vii) The government should establish a special system called the National Investment Fund (NIF) to facilitate the financing of long-term investment in plants and equipment for the industries. The sources of the NIF should be a combination of domestic funds from private financial intermediaries such as commercial banks and government. The NIF should mainly supply equipment loans to facilitate construction of the industries, specifically such heavy industries like steel and composites, automobiles, aircraft, petrochemical, and shipbuilding industries.

(viii) Nigerian government must overhaul the education and training systems to promote and secure engineers and skilled workers for the industries. There should be update of training centers, technical high schools, and engineering colleges. Specifically, the government should impose vocational training requirements on private sector firms to expand the supply of skilled labor for the industries.

(ix) The government should also introduce a skills licensing system to encourage every Nigerian worker to possess at least one skill. In addition, for each field of engineering the government should actively recruit outstanding Nigerian scientists abroad and establish a modern laboratory where research on the improvement of production technologies will be conducted in collaboration with industry researchers and university professors.

(x) There should be risk sharing between the Nigerian government and private firms to promote the process of rapid industrialization and product diversification.

(xi) The Nigerian government should act as an active risk partner for all industrial firms chosen to participate in strategic projects. In practice, the risk-sharing scheme will be established by the state's control over finance. The government's commitment to risk partnership will largely motivate private entrepreneurship and allow the credit-based economy and its highly leveraged firms to explore risky investment opportunities with long-term objectives in mind.

(xii) All major banks should set their interest rates at levels far below market rates, and tightly control the allocation of their loans and foreign loans.

(xiii) Large inflows of foreign capital should be promoted without political interruption, and risky ventures that could not be undertaken by private companies alone could be undertaken with government support. Furthermore, government, by controlling financial markets, must not hesitate to bail out whatever strategic firms were financially insolvent.

(xiv) To prevent financial crises occurring among exporting firms, risk-sharing schemes between creditors and borrowers should be backed by government's direct involvement in risk sharing with business firms through a constitutional act termed "the Presidential Emergency Decree" which should declare a moratorium on the payments of corporate debt.

(xv) Government should ensure that all corporate loans from the curb market were converted to long-term loans to be paid on an installment basis over a ten-year period with a grace period of four years, at a maximum interest rate of not more than 3%-5%.

(xvi) Strong market competition among private firms in Nigeria should be encouraged.

(xvii) Export promotion policies which always lead to substantial technological spillover should be highly promoted which will in turn stimulate learning by doing.

(xviii) The size of policy loans being estimated for earmark as credit facility to exporting firms such as export credits should be enforced, because there is some positive correlation between export growth and export loan support, in terms of its availability and the extent of preferential treatment.

## 4. Service Based Economic sustainable strategies (SBESS) – 2037-2040

Service industries are commonly known as tertiary industries. The term "services" covers a huge range of economic activities, including retailing, banking, insurance, catering, medicine, law, accountancy, cleaning, teaching, television production, the civil service, sport, transport, and many more activities.

Over the past century the service sector has expanded in the developed world. The service sector is now the most important sector in the advanced economies, accounting for about two thirds of the total economy in countries such as the United Kingdom and the United States.

The retailer deals directly with consumers and must be aware of and even anticipate their needs and desires. Some of the larger retail firms are the discount store, chain store, department store, and supermarket. Retailing also includes house-to-house canvassing, mail-order selling, vending machines, petrol stations, and street stalls.

Nigeria's projected industrialization phase and future pace must be determine by her seriousness, aggressive pursuit and resilience. A fully industrialized Nigeria, must encourage a rapid growth and development in its service based sector. This must be in line with standard global competitiveness. Present up coming and growing retail giants like Jumia and Konga must stand up and muzzle themselves to be able to compete with their advanced counter parts overseas like Amazon, EBay, McDonalds etc so they can transition along with the country to an advanced service based tertiary economy after Nigeria must have completely industrialize some decades from now. In the future, the Nigerian service based sector should be able to :-

(i) Contribute minimum 70-85% GDP to the Nigerian economy.

(ii) Employ large number of employees beyond the Federal, States and Local governments put together.

(iii) Use advanced technology in package delivery in the shortest possible time especially the use of civilian delivery drones to buyers' doorstep.

(iv) Command large number of patronage and followers both locally and internationally and compete favorably with overseas sister corporations.

(v) Use and prioritize advanced digitalize service provision across all boards and maintain a functional central data in a Cloud based superior super AI computer.

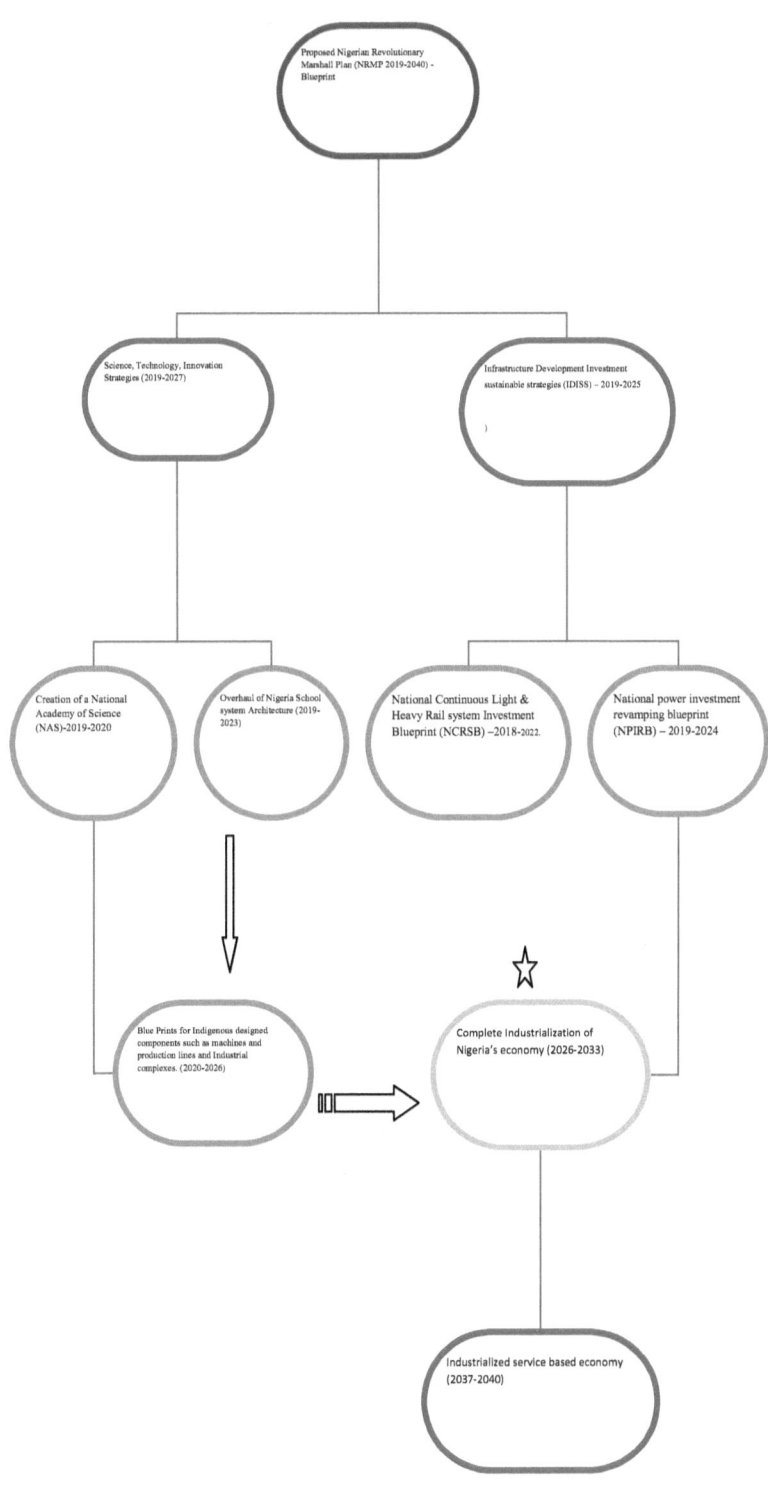

Graphical representation of the proposed Nigerian revolutionary Marshall plan (NRMP) 2019-2040 - blueprint.(above)

### 4.1.3 Re-design of Nigeria's Education system architecture

Introduction

Education denotes the methods by which a society hands down from one generation to the next its knowledge, culture, and values. The individual being educated develops physically, mentally, emotionally, morally, and socially. The work of education may be accomplished by an individual teacher, the family, a church, or any other group in society. Formal education is usually carried out by a school, an agency that employs men and women who are professionally trained for this task.

The educational systems in the countries of the Western world were based on the religious tradition of the Jews, both in the original form and in the version modified by Christianity. A second tradition was derived from education in ancient Greece, where Socrates, Plato, Aristotle, and Isocrates were the influential thinkers on education. The Greek aim was to prepare intellectually well-rounded young people to take leading roles in the activities of the State and of society.

# Historical Background Influence to Nigerian Educational System

The oldest known systems of education in history had two characteristics in common: they taught religion, and they promoted the traditions of the people. In ancient Egypt the temple schools taught not only religion but also the principles of writing, the sciences, mathematics, and architecture.

The Bible and the Talmud are the basic sources of information about the aims and methods of education among the ancient Jews. Jewish parents were urged by the Talmud to teach their children such subjects as vocational knowledge, swimming, and a foreign language. Today, religion serves as the basis for education in the home, the synagogue, and the school. The Torah remains the foundation of Jewish education.

As the Roman Empire declined, Christianity became a potent force in the countries of the Mediterranean region and in several other areas in Europe. The earliest types of Christian education were the catechumenal, or neophyte, schools for converts; the more advanced catechetical, or question-and-answer, schools for Christians; and the Episcopal, or cathedral, schools that trained priests. The early Fathers of the Church, especially St Augustine, wrote on educational questions in light of the newly adopted Christian concepts. Many monasteries or monastic schools as well as municipal and cathedral schools were founded during the centuries of early Christian influence.

Collections, or compendiums, of knowledge centred on the seven liberal arts: the trivium, composed of grammar, rhetoric, and logic, and the quadrivium of arithmetic, geometry, astronomy, and music. From the 5th to the 7th century these compendiums were prepared in the form of textbooks by such scholars as the Latin writer Martianus Capella from northern Africa, the Roman historian Cassiodorus, and the Spanish ecclesiastic St Isidore of Seville. Generally, however, such works disseminated existing knowledge rather than introducing new knowledge. The Protestant Churches deriving from the Reformation instituted by Martin Luther in the early 16th century established schools to teach reading, writing, arithmetic, and catechism on the elementary level; and the classical subjects, such as Hebrew, mathematics, and science, on the secondary level. Public education grew in the 19th century from nationalistic, as well as religious, motivation.

Universal elementary education was difficult to achieve without compulsion. Laws that made school attendance compulsory were passed in Massachusetts in 1851; also significant in the 19th century was the widespread organization of missionary education in the undeveloped areas of the world, particularly in Africa and Oceania. Education in colonial areas such as India was given attention by the administrative powers. In general, however, the vast majority of the people in the colonial and underdeveloped regions received little, if any, formal education. Progressive education was a system of teaching based on the needs and potentials of the child, rather than on the needs of society or the precepts of religion. The 20th century was marked by the expansion of the educational systems of the industrial nations, as well as by the emergence of school systems among the newer, developing nations in Asia and Africa. Compulsory basic education has become nearly universal, but evidence indicates that large numbers of children, perhaps 50 per cent of those of school age throughout the world, are not attending school. In order to improve education on all levels, the United Nations Educational, Scientific and Cultural Organization (UNESCO) inaugurated literacy campaigns and other educational projects. The aim of this organization is to put every child everywhere into school and to eliminate illiteracy. Some progress has been noted, but it has become obvious that considerable time and effort are needed to produce universal literacy.

## Nigeria's Education Background

Traditional Koranic schools are widespread throughout the north, and missionaries brought Western education to the coastal areas as early as the 1830s. Until the 1970s, enrolment in Western-oriented schools was significantly higher in the south. In 1976 free primary education was established throughout Nigeria. Educational facilities are insufficient, however, and the adult illiteracy rate remains about 57 per cent. In 1994, some 16.1 million pupils were enrolled in primary schools and more than 4.8 million students attended secondary and tertiary schools. Under a new educational system introduced in 1982, primary schooling (officially compulsory) takes six years to complete. Secondary schooling is organized in two successive phases of three years each. In 1994, 1.7 per cent of the country's gross national product (GNP) was spent on education. Western-style higher education, begun in 1948 with the founding of the University of Ibadan, is found throughout the country. There are 31 universities and more than 380,000 students attend 133 higher education institutions. Other major institutions include Ahmadu Bello University (1962), in Zaria; the Obafemi Awolowo University (1961), in Ife; the University of Lagos (1962); and the University of Nigeria (1960), in Nsukka. British-style universities have been augmented by a growing system of American-influenced teachers' colleges and technical colleges.

The objectives of the Nigeria 1981 National Policy on Education is to train manpower in Engineering, Applied Science, Technology and Commerce at all professional grades, and the provision of technical knowledge and vocational skills necessary for agricultural, industrial, commercial and economic development. These objectives have become mere theories and have remained on the pages of papers. A review of Nigerian Polytechnic systems reveals that Polytechnic education is underrated, discriminated against, disowned, neglected, denigrated, snubbed and ignored. The standard of education is falling and the consequence is the production of half-backed graduates and job seekers who incidentally roam the streets looking for white collar jobs.

Ibiyemi (2007) stressed that Nigerian educational system has been disorganized. He feels that university education is not for everybody, though everybody should be given a chance to prove their ability.

We should devote all our efforts to ensure that Nigeria invests adequately in the training of our children locally and in science so that they are in a position to fully appreciate science and technology and its capabilities. Our young children should be made to know quite early in school that science and technology plays a vital role in determining the efficacy of a country's economy and the way people live. Most people know about or have used such technology as television, motor vehicle, telephone, iPad, printer, aeroplane, computer, the Internet, email and MRI used in diagnostic medicine. The list goes on. The point is that today technology is so ubiquitous that there is virtually no facet of our lives that is not impacted by the beneficial effects of technology whether we know it or not. It is reasonable to begin early to expose our children to the culture of science and to show them how technology is used to address specific societal problems. Since the state of most of our schools is less than desirable, it goes without saying that Nigeria has a moral duty to upgrade our schools with well-equipped and up-to-date facilities that will promote education at the elementary, secondary, university and tertiary levels where the teaching of science should be encouraged and given the priority it deserves. By so doing Nigeria will sow the seeds for the technological revolution that the country badly needs. In the case of university research, funding should be competition-based, to provide incentives to enhance efficiency and strengthen collaboration with industry.

Mechanisms must be put in place to improve the quality of equipment and facilities available for teaching at all levels, as well as for research at the tertiary level. These measures are not limited to simply increasing funds available; collaboration between higher institutions (national and international), a more limited focus in research programs offered, and an improvement in the culture of maintenance are all measures that will help.

The Nigerian government is at a crossroads. Either we continue to decline, or we focus on improving our schools so that we build the workforce required to become a nation of makers. Moving forward, the nation must focus on improving students' reading skills at an early age. Reading is the foundational skill without which educational achievement, and the social and economic advancement that it makes possible, becomes unattainable. One of its key initiatives, manufacturing a desirable career path, is designed to help schools and colleges graduate on-time literate and skilled career-ready citizens who are equipped to join the workforce or continue in postsecondary education.

At the end of World War II, America had become the world's

leading manufacturing powerhouse, producing 50% of the world's industrial output. America was then the leader in quantity and quality of high school diplomas. Recent developments point to a growing awareness of the importance of Career Technical Education (CTE) programs "CTE is an important pathway for students to prepare for the workforce by integrating practical applied purposes with work-based knowledge and a hands-on learning experience." to equip educators with tools designed to inspire and teach CTE, advanced manufacturing careers and Science, Technology, Engineering and Mathematics (STEM), and more. Manufacturing companies can work with partners in the community to create a stronger pool of prospective candidates. Executives indicate that engaging early with local schools and colleges in science, technology, engineering, and mathematics (STEM) initiatives and skills certification programs can close the skills gap. This change is significant. The nation is now facing a critical "skills gap" between the number of talented professionals available and the number needed to lead manufacturing and design in Nigeria. Eighty-two percent (82%) of manufacturers surveyed by the Manufacturing Institute and Deloitte reported a moderate or serious shortage of skilled workers. By 2020, 65 percent of all jobs in the economy will require postsecondary education and training. Nigeria needs a National Assessment of Educational Progress (NAEP) commission. Low literacy levels cost the U.S. $225 billion or more each year in non-productivity in the workforce, in crime, and in loss of tax revenue due to unemployment. Superimposed on these have been influences of British colonial rule and European missionary educational systems. During the 1970s an increasingly self-confident federal government sought to modernize Nigeria rapidly, using Western-style education as a major tool. Revenue from the sale of crude petroleum helped to finance such modernization.

### 4.1.3.1. Proposed Recommendations on education (Recommendations for Restructuring / Updating of Nigerian Education system Architecture)

Proposed Nigeria Schools Reform Act (NSRA) - 2019-2023. A three (4) years aggressive reform and restructuring of all Nigerian schools should be carried out. A legislative act of the national assembly school be passed and signed into law to give the process a legal backing and proper starting point.

1. Horizontal Education System (HES); This entails a science and technology oriented horizontal system of education where a student should pursue a specific science and technical discipline from primary 5 to the university undergraduate and post-graduate levels to enable him/ her master the particular field and become a consultant and/or researcher in the field of discipline latter in life. In the horizontal education system, a particular discipline only gets advanced whenever a student or candidate crosses from one level to the other example

(a) Primary 5 – 6  (introduction to general science.)

(b) JSS 1 – 3 (to be taught core science and technical subjects and be encouraged through advice from guidance and counseling to make their choice of science and technical profession).

(c) SS1 – 3 (to be taught the rudiment of their chosen profession e.g. Applied Physics, Biochemistry, Petroleum engineering, building engineering, medical doctors, Nursing, Aeronautic engineering, Automobile engineer, structural engineer, etc )

(d) University Under-graduate / Polytechnic Diploma – HND level ( to be fully trained as professional engineers, scientists and technologists in their chosen career discipline capable of industry practice anywhere in the world)

(e) Post Graduate Diplomas (to be introduced to the particular discipline research pattern and engage in early team research duty)

(f) Post Graduate level in Master, M.Phil and P.hD ( to be full research fellows, project scientists and technologists involved in individual and specific group project and to be automatic bench members of the national academy of science).

2. Career Technical Education (CTE); At present in the country, it was observed that students applying and studying humanities in the arts and social sciences have far overshadowed those applying and studying the core sciences by margin of one to seventeen. This means that one seventeenth of every Nigerian student study a core science discipline. The career technical

education is meant to correct such gross imbalance and provide Nigerians with first rated professions like IT consultant experts, computer analysts and programmers, Industrial plant operators, Smartphone technicians, Draughts men, auto technicians, Airplanes technicians, Electronic experts, Acoustics specialists etc. Here, it means that every Nigerian irrespective of the previous course studied, must be made to go back and enroll in the nation's technical institutions like the polytechnic and other technical colleges and be trained to possess at least one technical career discipline that will enable him or her to work in the productive industrial sector. This will make Nigerians self reliant and self sustainable taking them away from the labor search market in search of white collar job but to be employers of labor themselves.

3. Advanced manufacturing careers (AMC); As Nigeria prepare to industrialize, a sizable number of its citizens must be trained in manufacturing and advanced manufacturing through

(a) theoretical and graphical presentation training,

(b) plant, machines and equipment simulation training and

(c) practical industry training overseas

this is to acquaint and familiarize the prospective industry career workers with industrial machines and plants, raw materials and finished products, production techniques and alternatives, production procedures and processes, industrial systems and economics, factory management, maintenance and consultancy and trading. This is a prerequisite to any industrialize society, so as to have a readily available workforce when the need arise and to take advantage of the vast population of the country for gainful productivity as witnessed in China, Russia and India. When industrialized, such industrial career workers will be rotated to different departments in shifts once in every two months to different industries to become specialized as against working in one plant their entire career and get only a limited knowledge.

4. Science, Technology, Engineering and Mathematics (STEM); scientific and technological revolution cannot take place in Nigeria except with the adoption of STEM in all aspect of its education. This will guarantee a future for the country and its populace. When all science, technology, engineering and mathematics intellects in the country are mobilized and empowered, they could set the revolution train rolling and bring about scientific and technological renaissance that would be felt everywhere in the country and abroad. STEM is a must for a country like Nigeria that needs holistic transformation at this period of its life cycle. STEM must be encouraged, advertised, publicized, inculcated and included in the curriculum of educational institutions even from the primary school level. No nation on earth can thrive without STEM neither can Nigeria be able to achieve scientific revolution nor attain industrialization.

5. Skills certification programs (SCP); Technical colleges and special technical centers should be provided and those available should be upgraded with modern state of the art equipments and

facilities. These citadels of learning located all around the country should set the pace in championing the acquisition of skills through initiation of strategic programmes capable of transforming even the most unattractive people into capable technical and skillful hands. Different skills in various fields of endeavor would be taught to near perfection like sewing, catering, welding, Aluminums partitions, doors and windows, carpentry and joinery, upholstery, interior décor, painting,

## 4.1.4. Proposed creation of ICSF:

## (Proposed Sci-Tech, R & D and Industrialization Funding – Independent consolidated strategic Fund (ICSF))

The issue of funding in any change mantra or agenda can never be overemphasized or exhausted. Suffice it to say that in such a giant leap to industrialize a large nation like Nigeria, a massive amount of funding is necessarily required. The first step is to prepare and present a deliberate bill for an act to create an "Independent consolidated strategic fund (ICSF)" which will be responsible for funding science and technology (S&T), research and development (R&D), and industrialization, and legislate upon it at the national assembly. When successful, and it became a law signed by the president, it should immediately swing into action. The chairman of the Independent consolidated strategic fund (ICSF) should be a person of proven integrity, a retired or serving technocrat, transparent, incorruptible, nonpartisan and educated to PhD level, tested and trusted in all areas of life. To select the right qualified person that will be saddled with such a herculean national task, the following procedures must be satisfied according to the Act-

(a) there shall be a general meeting of all the chief executives of the three tiers of government( Federal, States and Local governments), the senate president, the speaker of the house of representatives and all speakers of states houses of assembly in the 36 states of the federation to come up with and inaugurate a high level and high powered tripatheid technical committee(TTC).

(b) The high level and high powered tripatheid technical committee (TTC) shall comprise of the following members

(i) chairmen of the traditional council of states in six geo-political zones of Nigeria,

(ii) The presidents of Nigeria Labor Congress (NLC) and trade union congress (TUC).

(iii) The chief justice of Nigeria and the president of the court of Appeal.

(iv) The presidents of National association of Nigerian students (NANS) and Nigeria youth council (NYC).

(v) The Presidents of Christian association of Nigeria (CAN) and Federation of Muslim societies (FMS)

(vi) The head of civil Service of the federation and one representative of head of civil service of the six geo-political zones of Nigeria.

(vii) Representatives of five (5) civil societies in Nigeria.

(viii) Chairmen of ten(10) major Nigerian political parties and representatives of association of Nigerian political parties.

(ix) Representatives (but nonvoting members) from UNCR, UNIDO, UNDP.

(c) The Tripatheid Technical Committee (TTC) shall select a chairman and three other members of the Independent consolidated strategic fund due to the importance attached to the body because it will be responsible for trillions of Naira in its account and it will be sorely looked upon to transit Nigeria to the next phase of its life cycle.

# The Proposed Independent Consolidated Strategic Fund and First Line Charge Account.

The funding of the proposed consolidated strategic fund shall be done through direct charge or first line charge on all proceeds accruing from the following

1. All sales of bye products of crude oil refining such as paraffin wax, tar, petroleum jelly, etc ;
2. 5% of all crude oil sales per barrel of exported crude oil.
3. 5% from the sale of liquefied natural gas (LNG).
4. 10% remittance to CSF account quarterly from the excess crude oil account.
5. From amounts saved from importation of raw materials (N3trillion naira in the next 5 years).

6. More effective partnerships between the consolidated strategic fund which is a public body and the private sector should be encouraged and established through private participation in the form of direct sponsorship or indirectly through the public private partnership (PPP).

This is because in developed countries, research and development is mainly driven by the private sector. Close to half of the world's research and development expenditure is accounted for by only 700 private firms, according to the OECD. Private companies contribute 76% of the gross expenditure on research and development in Japan; in the USA, they contribute 70%. In South Africa, they contribute 42.7%, whereas the government contributes only 36.4%. So leveraging the private sector is crucial; any existing barriers to this kind of partnership should be removed. But since Nigeria's economy is still developing, at this stage it is not expected to perform like those of the advanced economies. First Nigeria need to industrialize through sustainable effort of science, technology, research and development to do so, the government must be the principal drive actor of the transformational process rather than the private sector.

However, the following five key enablers are examples of some opportunities to financing science, technology, research and development and industrialization:

(a) Catalyze funding into infrastructure and industry projects that attract further investments by increase and channeling funding into GDP catalytic programmes.

(b) Improve access to market finance for Nigerian enterprises, advising governments, stock exchanges and regulators on development of liquid capital markets and Foster partnerships.

(c) Increase firm capabilities and generate important productivity spillovers (through technology and know-how transfers, financing can lead to higher productivity jobs and high-value added industry niches);

(d) Access to markets, in particular small domestic markets can drive enterprise development and scale- up investments and financing to SMEs and establish linkages of SMEs to domestic projects/companies;

(e) Foster successful industrial policies that facilitate spillovers and backward linkages, and incentivize key PPP projects.

These enablers will require vision and commitment from political leaders but also from Financing Institutions, private sector and the broader development community, to provide support through technical assistance, capacity building, continuous dialogue, partnership and advisory services.

Overall, commitments to support Africa's and indeed Nigeria's economic structural transformation have never been stronger with regard to the adopted SDG 9. These commitments put world leaders and partners at a center stage for operationalizing Africa's industrialization. The time has come to transform these good intentions into tangible actions by initiating a progressive collaboration between the United Nations Industrial

Development Organization (UNIDO), the Office of the Special Adviser on Africa (OSAA) and the African Union, this will provide an opportunity for key stakeholders and development partners to address the topic and find a lasting solution to the issue.

### 4.1.4.1. Beyond Politics (Overcoming the Barrier)

(i) Programme continuity and consistency; Policy continuity is lacking in Nigeria; according to Akuta (2009) once a new government takes over power, it usually abandons previous governmental policies or programs. Very few policies have been retained by subsequent administrations (both military and civilians). These policy inconsistencies are an avenue for siphoning public funds. Every government policy, program or project cancelled means loss of government revenue, it brings with it destabilization and sets the people and the nation backwards especially when people have invested money, time and other resources. Nigeria receives a large income from oil production and the export of natural gas to neighboring countries in West Africa. The lack of programme continuity has a significant impact on nearly all of the country's development goals.

Successive governments must ensure the continuity of the industrialization programme irrespective of partisan politics. The industrialization programme must be observed and sustained. If the National programme on industrialization becomes an act of the national assembly and is passed into law, and is enshrined in the country's constitution, then it has the necessary legal backing for continuity and sustainability. The Bill for an act to create the industrialization programme should include a clause for a review or an amendment to update the Act only after twenty five (25) years. This will prevent any bureaucratic bottle necks and unnecessary interference with the industrialization programme.

## (ii) Interference in activities of the Independent Consolidated Strategic Fund (ICSF)

The Bill that proposed the creation of the consolidated strategic fund (CSF) and established it into Law should include some clauses that will empower the fund and prevent any individuals or groups no matter how powerful from any intended form of interference or meddling in the structure or activities of the body. All hands must be on deck and all eyes watchful if we desire to move this nation, our nation forward. All respectable and patriotic Nigerians must get on board this ship and help steer the nation in the right direction. We cannot afford to do otherwise.

Any form of commission or omission intended to inhibit the smooth running of the body by any person or group of persons, organization or government authorities must be prevented through the Act which creates and empowers it (CSF) to be totally independent. The duration of the independent consolidated strategic fund shall be in tandem with that of the industrialization Blueprint as the former ensures the sustenance of the programme. In the event of the indisposition of the chair of the body due to ill health, old age or dead, the secretary of the body shall take charge and be sworn in as the substantial chairman. The national industrialization programme shall be a long term goal as outlined in the blueprint and shall not be less than 30 to 50 years from the date of commencement.

# CHAPTER 5:

## RECCOMMENDATIONS FOR NIGERIA'S PERMANENT ALERT POSTURE

### 5.1. Enforcing multi-dimensional economic strategies.

(1) There should be a strong export oriented economic policy framework.
(2) There should be strong domestic demand.
(3) There should be strong retail sales.
(4) It should encourage industrialized economy and not just a sector of the economy or a government MDA.
(5) There should be a strong foreign developed consumption partner.
(6) There should be modest corporate and household debts.
(7) There should be heavy investment in infrastructure.
(8) There should be government fiscal stimulus measures.
(9) There should be equal level of income distribution.
(10) There should be a defined area of specific competitive advantage.
(11) There should be development in the intellectual abilities of our human talent (that is education and training).
(12) Nigeria should promote technological, innovative, research and development and an export oriented industrialization.

(13) Nigeria should celebrate scientists, technologists, researchers, entrepreneurs, industrialists and developers who have superior intellectual abilities and contribute directly to the rapid transformation of every country's economy over and above other intellectual properties who do not contribute directly to nation building.

(14) All Nigerian roadmap and other postulated developmental plan should be a strategic document whose target and end goal is to rapidly project Nigeria to become an advanced and high-income industrialized developed country.

(15) Just like socialism is a movement towards finally becoming communism, so should Nigeria being a mixed economy strive towards becoming a service based tertiary economy which could command a GDP of between 65-90% to the economy.

(16) There should be a strong drive for technology transfer through training and retraining of scientists, technologists and other professionals overseas to school and work in factories and industries so they can acquire skills and acquaint themselves with blue prints and operating principles of those industrial plants, equipments and machines which they will later come back and initiate the design and fabrication of these modified indigenous products.

## 5.2. Encouraging Home/garage Laboratories

## (Create & organize small research Teams- In the name of patent)

(a) Home managed "garage" labs:

The Industrial Revolution, which occurred from the eighteenth to nineteenth centuries, hundreds of years, was a period amid which overwhelmingly agrarian, rustic social orders in Europe and America got to be mechanical and urban. Before the Industrial Revolution, which started in Britain in the late 1700s, assembling was frequently done in individuals' homes, utilizing hand devices or essential machines.

Nigerians at all levels must be encouraged to start their own home laboratory, that was how the industrial revolution began in Europe at homes, inside stores and garages or at backyards through the individual efforts of some few committed scientists and innovators. Today, so many individual scientists and innovators of high repute have converted some facilities in their personal residences to become mini research centers or so called "garage labs" at home and dedicated same to testing and assembling of new technologies. Such humble beginning from the home garage laboratories usually turns out to produce some of the most important inventions and innovations in human history which leads to several patent claims and awards. Invention is about creating something new, while innovation introduces the concept of "use" of an idea or method. . An

invention is usually a "thing", while an innovation is usually an invention that causes change in behavior or interactions. We don't want to downplay the importance of
invention. Documenting, protecting, and leveraging inventions is a cornerstone of innovation.

Man has an unquenchable innate curiosity and has continually tried to find out through scientific research how the natural world works. Scientific research can be classified under three headings: (a) basic or fundamental research, (b) applied research, and (c) developmental research. Basic research is the search for truth about nature for its own sake. The sole aim of basic research is to contribute to the pool of new knowledge and thereby provide a better understanding of the subject being studied. It is the search for truth for its own sake, and its intent is not to create or invent a product. Basic research is very expensive and is usually funded by the government. In contrast to basic research, applied research is usually carried out to address a specific problem and it leads to products or services or to solutions to important problems that face society. Applied research is usually funded and carried out by companies, the government, research institutes and the universities. In the process of carrying out applied research important new basic information about nature may be uncovered. So the distinction between basic and applied research cannot be made too rigid. Finally, research aimed at turning applied research discoveries into large-scale, marketable commercial concerns is developmental research. In Nigeria this quintessential continuum from basic to applied to developmental research is, as yet, practically nonexistent. All Nigerian scientists, technologists and researchers in all fields of human endeavors are through this medium encouraged to:-

(a) dedicate a safe room or a section of their apartment as a science and technical lab where it is always accessible to them when home and away from their official duty post. This is essential because some sudden unexpected ideas on a proposed postulated theorem or innovation may emerge or dawn on the scientists or inventors while home or in a social gathering for the furtherance of his or her work.

(b) Home or garage laboratories provide such a serene atmosphere which scientists, technologists and inventors needed to concentrate on their work without undue distractions.

(c) Home or garage laboratories will always serve as an assessable platform or medium which must be constantly and readily available for accessibility whenever it is needed.

(d) Home or garage laboratories should serve as a secure diversion away from copy carts and other forms of corporate espionage which mostly targeted institutions, corporations and government research outfits.

(b) Organized small formal home/garage research Teams:

Isaac Asimov (the science fiction writer), recommends

"cerebration sessions" to promote innovation beyond invention, he said shared thinking and informal collaboration as a key component of problem solving and accelerating change. Organized research groups will command "Group think" new ideas, new possibilities, and new combinations of knowledge and experience which could find new answers and new directions. First and foremost, there must be ease, relaxation, and a general sense of permissiveness. But innovation comes from people meeting up in the hallways or calling each other at 10:30 at night with a new idea, or because they realized something that shoots holes in how we've been thinking about a problem. –Steve Jobs.

Eric Schmidt and Jonathan Rosenberg considered common behaviors and environments at some big companies considered successful in innovation such as Google, Apple, Samsung, etc which has the culture of cerebration, of promoting employees to share and internally market their ideas and projects. Some of this could be influenced by a higher concentration of like-minded talent, such as in Silicon Valley USA, where former colleagues and trusted friends have migrated from company to company while maintaining their social networks and friends of shared interests and hobbies. But all of these companies have also created a corporate culture that recognizes the value and the opportunity of cerebration and group-think. They have created a system that operates both as a business, and as an idea factory.

We learned that the only way for businesses to consistently succeed today is to attract smart creative employees and create an environment where they can thrive at scale. It's best to work in small teams, keep them crowded and foster serendipitous connections. We need to focus our research efforts and expenditure on the few areas where we have comparative advantage – areas like agriculture, petrochemicals, renewable energy, and mining.

(c) Home/Garage laboratories- Encouraged and motivated by Patent (even breakthroughs of the Industrial revolution).

Patents are evidence of inventions, of having thought of something first, and documenting the new invention through a legal process. The usefulness of those inventions is not proven, so "inventions" do not always equate to "innovations." There are many patents which really do not have a use or have influenced no products or industries. Patents without a "use" are not innovation. If innovations infer the "use" of a new idea or method, then an invention that leads to innovation is really qualified by how much it changes the behaviors of the users, the businesses, and the processes around it. Now perhaps the "Nose Pick" patent was a victim of bad marketing, poor manufacturing, or just a "right idea at the wrong time", but obviously it has not changed behavior and become a commonplace item. Nowadays, some companies often claim to be a "leader in innovation", and show a large pile of patents as evidence.

The patent process and legal systems around the world recognize the rights of an inventor and help them by establishing a system which allows them the opportunity to exploit their inventions for financial gain for a given period of time. Invention rights owners can produce products without others blatantly copying, license their inventions to others to produce, or create combinations of inventions by partnering with other intellectual property owners.

Inventions of the Industrial Revolution that Changed the World are generally encouraged and in some instances motivated by patent claims. The period between the mid 1700's and mid 1800's was one of great technological and social change. The Industrial Revolution was a period of rapid social and technological change that has shaped the world we live in today. It was a period of great innovation and many of the items we see today were inventions of the Industrial Revolution. They range from innovations within the textile industry to the iron industry and consumer goods of the later Industrial Revolution. The period between the mid-1700's and 1840 is **commonly agreed** to be the period of time for the Industrial Revolution. Here are some of the patented inventions of the industrial revolution that changed the world forever.

Flying shuttle or weaving made easy, was widely used throughout Lancashire after 1760 and was one of the key developments of the period. It was patented in 1733 by John Kay, and its implementation effectively doubled the output a weaver could make, thereby allowing the workforce to effectively be halved. Prior to this invention, a weaver was required on each side of a broad-cloth loom, now one weaver alone could do the job of two. Several subsequent improvements were made to it over the years with an important one in 1747. Its impact was incredibly significant, effectively allowing production of textiles beyond the capacity of the rest of the industry. It arguably prompted further industrialization throughout the textile and other industries to keep up.

**The Spinning Jenny increased wool mills productivity;** this was another example of great inventions of the Industrial Revolution. It was developed by James Hargreaves who patented his idea in 1764.The Spinning Jenny was groundbreaking during its time and one that would help change the world forever. It allowed workers to spin more wool at any one time. This vastly increased mills productivity and, along with the Flying Shuttle, helped force further industrialization of the textile industry in the United Kingdom. It allowed for a massive reduction in the work needed to produce a piece of cloth and allowed for a worker to work eight or more spools at a time. With further refinement, this increased to 120 spools over time. It has long been credited as the main driver for the development of a modern factory system. By the time of Hargreaves's death in 1778, there were around 20,000Spinning

Jennys across the UK.

**The Watt Steam Engine, the engine that changed the world.** When James Watt created the first reliable steam engine in 1775 his invention would literally change the world. His innovation blew the older less efficient models, like the Newcomen engine, out of the water. James' innovation of adding a separate condenser significantly improved steam engine efficiency, especially latent heat losses. His new engine would prove very popular and would wind up installed in mines and factories across the world. It was hands down, one of the greatest inventions of the Industrial Revolution. His version also integrated a crankshaft and gears and it became the prototype for all modern steam engines. It would eventually lead to incredible improvements in almost all industries, including the textile industry, across the world. Steam engines would also lead to the development of locomotives and massive leaps forward in ship propulsion.

**The Cotton Gin: the engine that made cotton production boom.** Eli Whitney is another name synonymous with inventions of the Industrial Revolution. He invented the cotton engine, gin for short, in 1794. Prior to its introduction into the textile industry, cotton seeds needed to be removed from fibres by hand. This was laborious and time-consuming to say the least. This machine vastly improved the profitability of cotton for farmers. The Cotton Gin enabled many more farmers to consider cotton as their main crop. This was especially important for farmers and plantation owners in the Americas. With the seeds and fibres separated more efficiently it became much easier for farmers to use the fibres to make cotton goods like linen. They could also simultaneously separate seeds for more crop growth or the production of cottonseed oil.

**Telegraph communications, a pillar of the Industrial Revolution**
Coming in at the tail end of the Industrial Revolution, the Telegraph was one of the greatest inventions of the Industrial Revolution. Created in the early 1800's it would change communication forever. Thanks to this technology, near instant communication became possible initially across the country and eventually across the globe. This enabled people to stay in contact and become aware of wider geopolitical events much more easily. The first true electrical telegraphs finally superseded optical semaphore telegraph systems to become the world's first electrical form of telecommunications. In only a matter of decades, the electrical telegraph became the de facto means of communication for business and private citizen's long distance.

Joseph Asp din was a bricklayer turned builder who, in 1824, devised and patented a chemical process for making Portland cement. This one invention from the Industrial Revolution has been one of the most important of all time for the construction industry. His process involved sintering a mixture of clay and limestone to around 1,400- degree centigrade. This then needed to be ground into a fine power only to be later mixed with sand and gravel to make concrete. Years later, Brunel would use Portland cement to help construct the Thames Tunnel. It was also used on a large scale in the construction of the London Sewage system and many other construction projects around the world.

Henry Bessemer in 1856 patented the Bessemer technique. **The Bessemer process that changed steel** production was the world's first inexpensive process for mass production of steel from molten pig iron. This would also prove to be one of the greatest inventions of the Industrial Revolution. It is noted for its removal of impurities from the iron via oxidation as air is blown through the molten metal. Oxidation also helps raise the temperature of the iron mass to keep it molten for longer. The process is named after its inventor. The ability to mass produce high-quality steel and iron allowed a literal boom in the use of them in many other aspects of the revolution. Iron and steel suddenly became essential materials and would be used to make almost everything from appliances to tools, machines, ships, buildings, and infrastructure.

The first modern Battery by Volta, although there is evidence of early batteries from the **Parthian Empire** around 2,000 years ago, the first true modern electric battery was invented in 1800. This world first was the brainchild of one Alessandro Volta with the development of his voltaic pile. Mass production of the world's first battery began in 1802 by William Cruickshank. The first rechargeable battery was invented in 1859 by the French physician Gaston Plante. Later advancements would lead to the Nickel-Cadmium battery being developed in 1899 by Waldemar Junger. Volta's initial invention literally sparked a great amount of scientific excitement around the world which would lead to the eventual development of the field of electrochemistry.

**The Locomotive revolution is** the invention of the steam engine which would eventually lead to a revolution in transportation around the globe. **Locomotives** allowed large-scale movement of resources and people over long distances. Previously the industry relied on man- and animal- powered wagons and carts. After the pioneering work, Richard Trevithick in 1804 and of George Stephenson and his "Rocket" train networks would begin to spring up all over the United Kingdom and eventually the world. The first public railway opened in 1825 between Stockton and Darlington in England, UK. This would be the first of many railways and locomotives that would revolutionize the way business and private citizens transport their goods and themselves around.

The first documented **factory** was opened by John Lombe in Derby around 1721. Lombe was granted a 14-year patent for his factory which used water power to help the factory mass produce silk products. The factory was built on an island on the River Derwent in the English county of Derby. The idea for the factory came to Lombe after he had toured Italy looking at silk throwing machines. On his return to the UK, he employed the services of the architect George Sorocold to design and build his new "Factory". Once completed the mill, at its height, employed around 300 people. On its completion it was the first successful silk throwing mill in England and, it is believed, the first fully mechanized factory in the world.

**The Power Loom, overtaking all UK factories.** The invention of the Power Loom effectively increased the output of a worker by over a factor of 40. It was one of the most important inventions of the Industrial Revolution. It was introduced in 1874 by Edmund Cartwright who built the very first working machine in 1785. Over the next 47 years, the Power Loom was refined until it was made completely automated by Kenworthy and Bullough. By 1850 there was around 260,000 Power Looms installed in factories all over the United Kingdom. Cartwright's power loom was first licensed by Grimshaw of Manchester who built a small steam-powered weaving factory in 1790. Sadly this soon burnt down. Initially, his looms were not a commercial success as they needed to be stopped to dress the warp. This was soon addressed over the next few decades as he modified the design into a more reliable automated machine.

Arkwright's Water Frame spinning machine. Richard Arkwright was a barber and wig maker who managed to devise a machine that could spin cotton fibres into yarn or thread very quickly and easily. In 1760 he and John Kay managed to produce a working machine. This prototype could spin four strands of cotton at the same time. He would later patent his design in 1769. Further refinement of his design would ultimately allow the machine to spin 100's of strands at one time. The spinning machine would go on to be installed in mills around Derbyshire and Lancashire where they were powered by waterwheels hence they were called water frames. Arkwright's machines alleviated the need for highly skills operators adding significant cost savings to mills that installed them.

**The Spinning Mule: the yarn game-changer** The Spinning Mule combines features of two earlier Industrial Revolution inventions: the Spinning Jenny and above-mentioned Water Frame. **The Mule** managed to produce a strong, fine and soft yarn that could be used in many kinds of textiles. It was, however, best suited for the production of muslins. The Mule was devised by Samuel Crompton in 1775 who were a too poor to actually patent his invention and so sold it to a Bolton manufacturer. The very first Mules were hand-operated but by the 1790's larger versions were driven by steam engines. These larger machines had as many as 400 spindles. The Spinning Mule would become a very popular machine indeed and was installed in a large number of factories but as he had relinquished his rights to the machine, Crompton would see none of the proceeds from the sales.

**Henry Cort's puddling process.** In 1784, **Henry Cort** succeeded in developing a method of converting pig iron into wrought iron by heating it and frequently stirring it in the presence of oxidizing substances. It was, at the time, the first method that allowed wrought ironed to be produced on a large scale. Henry had managed to save a large amount of capital during his 10-year service in the Royal Navy. With this money had bought an ironworks near Portsmouth in 1775. By 1783 he managed to obtain a patent for grooved rollers that would allow him to produce iron bars more quickly than the old method of hammering. His puddling process would take the iron industry by storm and over the following 20-years, British iron production quadrupled!

In 1792 William Murdoch **developed and introduced Gas lighting** in his house in Redruth, Cornwall, **Lighting the streets of the modern world.** These early gas lights used coal gas which was installed as lighting. Over a decade later German inventor Freidrich Winzer became the first person to patent the use of coal gas for lighting in 1804. A thermo-lamp was also developed in 1799 using gas distilled from wood and David Melville received the first patent in the U.S. for gas lighting in 1810. After its development, gas lighting became the method of street lighting across the United States and Europe. These would eventually be replaced with low-pressure sodium or high-pressure mercury lighting in the 1930's.

**2,000 cells to create the first Arc Lamp, Sir Humphrey Davy** was able to build the world's first arc lamp in 1807. His device used a battery of 2,000 cells to create a 100mm arc between two charcoal sticks. As impressive as his initial success, it was not a practical piece of equipment until the development of electrical generators in the 1870's. Arc lamps are still in use today in applications like searchlights, large film projectors, and floodlights. The term is usually limited to lamps with an air gap between consumable carbon electrodes. But fluorescent and other electric discharge lamps generate light from arcs in gas-filled tubes. Some ultraviolet lamps are of the arc type.

The Tin Can, jumping to new production heights was patented by a British merchant Peter Durand in 1810. It would have an incalculable impact on food preservation and transportation right up to the present day. John Hall and Bryan Dorkin would open the very first commercial canning factory in England in 1813. In 1846, Henry Evans invented the machine that can manufacture tin cans at a rate of sixty per hour. This was a significant increase over the previous rate of only six per hour. The very first tin cans had very thick walls and needed to be opened using a hammer. Over time they became thinner enabling the later invention of a dedicated can opener in 1858. It took the American Civil War to inspire the creation of tin cans with a key can opener as can still be found on sardine cans.

The Spectrometer revolutionizes how we studied glowing objects. In 1814, a German inventor, Joseph von Fraunhofer invented the spectrometer. His early device was devised to enable the chemical analysis of glowing objects. Little did Joseph know the full impact his invention would have on the scientific world? We can owe the fact that we know what the Sun is made of thanks to Fraunhofer. Thanks to Fraunhofer's contributions, Bavaria overtook England as the leader in optics research. He invented the spectroscope in 1814. In fact, his discoveries earned him a knighthood in 1824, two years before his death. Like all glassmakers of the time, he died early because of heavy metal poisoning.

Camera Obscura: The first photograph. Beginning in 1814, Joseph Nicéphore Niépce started a journey of discovery that would eventually lead him to become the first person to ever take a photograph. He would eventually do this using his new-fangled camera obscura that was set up in the windows of his home in France. The entire exposure took around 8 hours to capture the image. Joseph constructed his first camera in around 1816 which allowed him to create an image on white paper. But he was unable to fix it. He would continue his experimentation using different cameras and chemical combinations for the next 10 years or so. In 1827 he successfully produced the first, long-lasting image using a plate coated with bitumen. This was then washed in a solvent and placed over a box of iodine to produce a plate with light and dark qualities.

The first Electromagnet findings; the electromagnet was the culmination of a series of developments from Hans Christian Oersted, Andre-Marie Ampere, and Dominique Francois Jean Arago made their critical discoveries on electromagnetism. One man, William Sturgeon, would take the findings of these great scientists and build on them to build the world's first electromagnet. He found that leaving some iron inside a coil of wire would vastly increase the magnetic field created. He also realized that by bending the iron into a u-shape allowed the poles to come closer together, thereby concentrating the field lines. His design was improved upon by Joseph Henry who built, in 1832, a very strong electromagnet that was able to lift 1630 kgs.

The Mackintosh Raincoat; Perhaps one of the most useful of all inventions during the Industrial Revolution was when, in 1823, Charles Mackintosh devised the **Mackintosh**. Prior to his invention clothing was waterproofed by using a coating of rubber. But rubber would become sticky and tacky during hot weather and extremely stiff during winter months. Charles, a Scottish Chemist, successfully cured this problem and patented a new method of using rubber to waterproof clothing. Initially, he created his new waterproof clothing at his family's textile factory. By 1843, Mackintosh had begun mass production of their clothes and merged with a larger clothing manufacturing company. His method of waterproofing is known to us today as vulcanization. This process allowed the rubber to maintain its shape and not become sticky during hot weather like natural rubber. Mackintosh's design also placed the rubber covering inside two pieces of fabric rather than covering one.

**Modern Friction Matches made possible with wood.** In 1826,

John Walker gave the world the first modern matches. Early attempts were made to make a match that produced ignition through friction by Francois Derosne in 1816. These were, however, crude and used a sulfur tipped match to scrape inside a tube coated with phosphorus. This was both inconvenient and unsafe. Waker was a Chemist and druggist from Stockton-on-Tees who developed a keen interest in trying to make fire as easily as possible. Chemical combinations were known that provided sudden ignition but what hadn't been finalized was a means of transmitting the flame to a slow-burning material like wood. When, quite by accident, a prepared match ignited by accident from friction on the hearth he at once knew he had found the answer. He immediately set about producing wooden splints or stick of cardboard and coating them with sulfur. He then added a tip with a mixture of a sulfide of antimony, chlorate of potash and gum. Camphor was added later to mask the smell of the sulfur once ignited.

**Every great writer's companion, the Typewriter;** It is widely accepted that in 1829, **William Austin Burt** patented the "first typewriter" which he termed a "Typographer". There were earlier machines similar in purpose; a notable example being Henry Mill's 1714 patent, but it appears to have never been capitalized upon. The Science Museum in London describes Burt's machine as the "the first writing mechanism whose invention was documented". Despite its apparent breaking of new ground, contemporary sources indicated that even when used by Burt the machine was slower than handwriting. This was because the typographer needed to use a dial rather than keys to select each character. This lack of efficiency improvement over handwriting ultimately sealed Burt's machine's doom. Both he and its promotor John D. Sheldon never found a buyer for the patent. The modern typewriter would ultimately be invented in 1867 by Christopher Sholes.

**The Dynamo powered by the Faraday principle.** Here's another great invention of the Industrial Revolution. The basic principles of electromagnetic generators were discovered in the early 1830's by **Michael Faraday.** Faraday noted that electromotive force is generated when an electrical conductor encircles a varying magnetic flux. This would later become known as Faraday's Law. Michael also built the first electromagnetic generator, the Faraday Disk. This was a type of homopolar generator that used a copper disc that rotated between the poles of a horseshoe magnet. The first true dynamo, based on Faraday's principle, was built in 1832 by Hippolyte Pixii, a French instrument maker. His device used a permanent magnet that was rotated using a crank.

**Blueprints from Herschel and Poitevin; John Herschel**, a British scientist and inventor succeeded in developing the process that was the direct precursor to what we now know as blueprints. John made improvements in photographic processes, particularly in inventing the cyanotype process and variations (such as the chrysotype), the precursors of the modern blueprint process in around 1839. This process enabled the production of a photograph on glass; he also experimented with some colour reproduction. It is also believed that he coined the term photography. It wasn't until 1861 that 'true' blueprints were developed by Alphonse Louis Poitevin, A French Chemist. He found that Ferro-gallate in gum is actually light sensitive. When exposed to light it turns into an insoluble permanent blue. He successfully postulated that a coating of this on paper or other material could be used to copy an image from another translucent document. Who would have thought that this was one of the inventions of the Industrial Revolution?

**The Hydrogen Fuel Cell**; the Hydrogen **Fuel Cell** was first documented in 1838 in a letter published in the December edition of *The London and Edinburgh Philosophical Magazine and Journal of Science*. The piece was written by a Welsh physicist and barrister William Grove. In it, he described his development of a crude fuel cell that combined sheet iron, copper and porcelain plates and a solution of sulfate of copper and dilute acid. In the same publication published a year later a German physicist Christain Freidrich Schonbein also discussed his crude fuel cell that he believed he had invented. His letter described how current was generated using hydrogen and oxygen dissolved in water. Grove sketched his design later in 1842, once again, for the same journal. Both of these used similar materials used to day's phosphoric acid fuel cells. Those inventions patented by the inventors and innovators that made and shape them are either still in use today or paved the way for considerable advances in industries around the world.

# REFERENCES

1. Flying Shuttle Loom, Weaver's Cottage Museum
   Source: *Betty Longbottom/Wikimedia Commons*

2. Model Spinning Jenny, Museum of Early Industrialisation, Wuppertal, Germany Source: *Markus Schweiß/Wikimedia Commons*

3. Cotton Gin at Eli Whitney Museum.
   Source: *Brighterorange/Wikimedia Commons*

4. Hasler Electrical Telegraph
   Source: *Hp.Baumeler/Wikimedia Commons*

5. Interior of the Thames Foot Tunnel, mid-19th century.
   Source: *Nichtbesserwisser/Wikimedia Commons*

6. Example of Macadam Road in around 1850.
   Source: *Sutter County Library/Wikimedia Commons*

7. Bessemer converter at former ironworks, Hogbo, Sweden. Source: *Calle Eklund/V-wolf/Wikimedia Commons*

8. Volta's electric battery in Como, Italy.
   Source: *GuidoB/Wikimedia Commons*

9. Replica of the "Rocket", Nuremberg Museum, Germany.
   Source: *Urmelbeauftragter/Wikimedia Commons*

10. Lombe's Manufactory/Mill, Derby circa 1770.
    Source: *ClemRutter/Wikimedia Commons*

11. A working example of Arkwright's water frame, Helmshore Mills Textile Museum.
    Source: *ClemRutter/Wikimedia Commons*

12. Spinning Mule at Quarry Bank Mill. Source: *Black Stripe/Wikimedia Commons*

13. Davy's Arc Lamp and battery
    Source: *Chetvorno/Wikimedia Commons*

14. Source: *Chris Potter/Wikimedia Commons*

15. Joseph von Fraunhofer demonstrating the spectroscope.
    Source: *Richard Wimmer/Wikimedia Commons*

16. World's first photograph by Joseph Niepce
    Source: *Joanjoc~commonswiki/Wikimedia Commons*

17. Sturgeon's electromagnet. Source: *Chetvorno/Wikimedia Commons*

18. The Mackintosh, Circa 1893. Source: *Cbaile19 /Wikimedia Commons*

19. Source: *Jef-Infojef/Wikimedia Commons*

20. Burt's Typographer. Source: *Flickr/Wikimedia Commons*

21. Hippolyte Pixii's Dynamo.
    Source: *DMahalko/Wikimedia Commons*

22. Source: *Adrian Michael/Wikimedia Commons*

23. *Josh Floyd,* Anthony James *advisor on energy, systems and societal futures at independent research and education organisation the Understandascope, and*

*founding partner of the Centre for Australian Foresight. Swinburne University of Technology*

24. PRE Working Paper Series. Washington, D.C.: World Bank. Cho, Yoon-Je, and Joon-Kyung Kim. 1993. Credit policies and industrialization of Korea. Paper presented at the World Bank Symposium on the Effectiveness of Credit Policy
25. Sengupta, Jati K. 1991. Rapid growth in NICs in Asia: Tests of new growth theory for Korea. Kyklos 44 (4): 561-79. . 1993. Growth in NICs in Asia: Some tests of new growth theory. Journal of Development Studies 29 (2): 342-51.
26. Kim, K. S., and S. R. Park. 1988. Productivity change and factor analysis in South Korean manufacturing (in Korean). Seoul:
27. Pyo, H. K., B. H. Gong, H. Y. Kwon, and E. J. Kim. 1993. Sources of industrial growth and productivity estimates in Korea (1970-1990) (in Korean). Seoul: Korea Economic Research Institute.
28. Olukayode O Adebile Federal Polytechnic, Ede, Osun State, Nigeria E-mail: oaadebile@yahoo.com,
29. Dahud K Shangodoyin University of Botswana, Botswana E-mail: shangodoyink@mopipi.ub.bw Received: April 14, 2011 Accepted: July 11, 2011 doi:10.5539/jsd.v4n4p152
30. www.ccsenet.org/jsd Journal of Sustainable Development Vol. 4, No. 4; August 2011 Published by Canadian Center of Science and Education.

31. Journal of Sustainable Development Vol. 4, No. 4; August 2011 154 ISSN 1913-9063 E-ISSN 1913-9071, www.ccsenet.org/jsd.
32. Akuta, C. V. (2009). Inconsistent Policies and High Rate of Abandoned Government Projects', "Support Option A4 Group", Leicester-UK. Available: http://briefsfromakuta.blogspot.com/.
33. Akwenuke, B. (2008). Vision 2020: Nigeria Needs $600bn Investment. [Online] Available: http://rsagroup.net /indogeria/2008/05/06/vision-2020-nigeria-needs-600bn-investment/
34. Compton, V. (2004). The Relationship between Science and Technology, a Discussion Document prepared for the New Zealand Ministry of Education Curriculum Project. Available: http://www.tki.org.nzcurriculum/whats_happening/index_e.php).
35. American Wind Energy Association. http://www.altenergy.org/renewables/ CHAMCO, USA. Garbage Electricity:
36. New Collegiate Dictionary (1977). Springfield, Mass.: G. & C. Merriam.
37. Wikipedia the Free online Encyclopedia (2010). [Online] Available: www.wik
38. Municipal Solid Waste Conversion to Energy. [Online] Available: http://www.chamco.net/garbage_electricity.html
39. Development, Security, and Cooperation (DSC) (2007). Mobilizing Science-Based Enterprises for Energy,

Water, and Medicines in Nigeria. Available: www.nap.edu/openbook.php?record_id=11997&page=1

40. Dike, V. E. (2005). Corruption in Nigeria: A New Paradigm for Effective Control, Africa Economic Analysis. Available: www.AfricaEconomicAnalysis.org

41. Dike, V. E. (2009). Addressing Youth Unemployment and Poverty in Nigeria: A Call For Action, Not Rhetoric, Journal of Sustainable Development in Africa, Vol. 11, No.3, ISSN: 1520-5509

42. Ekpiwhre, G. (2008). Utilising science and tech to drive Vision 2020, Punch News Paper Publications Excerpts of a speech delivered at Tinapa resort Calabar during the Diaspora Day in July.

43. Esho, M. A. F. (2008). A keynote address delivered at the opening ceremony of the 4th international conference of the schools of Science, Engineering and Environmental Technology, The Federal Polytechnic, Ede. Federal Republic of Nigeria (1981).

44. National Policy on Education, (Revised); Lagos, Nigeria NERC Press.

45. Ibiyemi, T. S. (2007). Rule of law and due process: A priority for sustainable development in Sciences, Engineering and Environmental Technolgy in Developing Nations' Being a lecture delivered at the 3rd international conference of NACONSEET held at Federal Polytechnic, Ede, Osun State, Nigeria.

46. Iredale, Robyn, Fei Guo and Santi Rozario (2003). Return Migration in the Asia Pacific. Available: www.scidev.net/ 08/12/2003.

47. Ogbu, O. (2004). Can Africa Develop Without Science and Technology?, The African Technology Policy Studies Network, P.O. Box 10081, 00100 General Post Office. [Online] Available: http://www.atpsnet.org

48. Petters, S. W. (2011). Nigeria-Geological background: [Online] Available: www.onlinenigeria.com/geology/

49. Science Daily (2009). Is Garbage The Solution To Tackling Climate Change? Wiley–Blackwell, Retrieved April 22, 2010. Available: http://www.sciencedaily.com-/releases/2009/09/090929100654.htm

50. Victor, E. D. (2009). Technical and Vocational Education: Key to Nigeria's Development. Available: www.nigeriavillagesquare.com/.../victor.../technical-and-vocational

51. Webster's Yunus, M. (1998). Alleviating Poverty Through Technology, Essays on Science and Society.

52. https://www.vanguardngr.com/2012/09/nigeria52-nigeria-badly-needs-a-revolution-in-science-and-technology/

53. Leach, Graeme. "Industry", Microsoft® Encarta® 2007 [CD]. Microsoft Corporation, 2006.
54. Benjamin K. Sovacool. How long will it take? Conceptualizing the temporal dynamics of energy transitions, *Energy Research & Social Science* (2016). DOI: 10.1016/j.erss.2015.12.020
55. Glenn Marshall, Association for Manufacturing Excellence (AME) at marsh8279@aol.com, Richard Long Ed.D., co-author Blueprint for a Literate Nation at richardlong1854@gmail.com, Richard Hinckley Ph.D.

President/CEO Center for Occupational Research & Development (CORD) at hinckley@cord.o

56. Adenikinju, A. F. (2005). 'African Imperatives in the New World Order: Country Case Study of the Manufacturing Sector in Nigeria', in O. E. Ogunkola and A. Bankole (eds) Nigeria's Imperatives in the New World Trade Order. Nairobi: African Economic Research Consortium and Ibadan: Trade Policy Research and Training Programme, 101–58.

57. Adeoti, J. O. (2010). 'Investment in Technology and Export Potential of Firms in Southwest Nigeria'. AERC Research Paper No.231. Nairobi: African Economic Research Consortium.

58. Adeoti, J. O., Odekunle, S. O., and Adeyinka, F. M. (2010). Tackling Innovation Deficit: An Analysis of University-Firm Interaction in Nigeria. Ibadan: Evergreen.

59. Bamiro, O. A. (1994). 'National Technology Policy for Development: The Role of Research and Development Institutions'. Paper Presented at the National Workshop on Technology Management, Policy and Planning, NISER, October, Ibadan, 18–21.

60. Bevan, D., Collier, P., and Gunning, J. W. (1999). The Political Economy of Poverty, Equity, and Growth: Nigeria and Indonesia. Washington, DC: World Bank and Oxford: Oxford University Press.

61. Chete, L. N., Adeoti, J. O., Adeyinka, F. M., and Ogundele, O. (2014). 'Industrial Development and Growth in Nigeria: Lessons and Challenges'. WIDER Working Paper 2014/019. Helsinki: UNU-WIDER.

62. Fashoyin, T., Matanmi, S., and Tawose, A. (1994). 'Reform Measures, Employment and Labour Market Processes in the Nigerian Economy: Empirical Findings', in T. Fashoyin (ed.) Economic Reform Policies and the Labour Market in Nigeria. Lagos: Friedrich Ebert Foundation, 5–13.

63. Federal Government of Nigeria (FGN) (2008). Yar' Adua's Seven Point Agenda. Abuja: Federal Ministry of Information and Communications.

64. Forrest, T. (1993). Politics and Economic Development in Nigeria. Oxford: Westview Press.

65. Manufacturers Association of Nigeria (MAN) (2009). 'Sustaining Nigeria's Manufacturing Sector in the Face of the Current Global Economic Recession'. Speech by the President of the Association, Alhaji Bashir Borodo, at its 37th Annual General Meeting, March, Lagos, available at <http://www.vanguardngr.co./2009/07/820-manufacturing-companies-close-down>, accessed 22 February 2016.

66. National Bureau of Statistics (NBS) (various years). Nigeria Gross Domestic Product Report. Abuja: Federal Republic of Nigeria.

67. National Planning Commission (NPC) (2004). Nigeria: National Economic Empowerment and Development Strategy. Abuja: National Planning Commission.

68. National Planning Commission (NPC) (2007). Nigeria: National Economic Empowerment Development Strategy (NEEDS2). Abuja: National Planning Commission.

69. (p.135) National Planning Commission (NPC) (2009). Nigeria Vision 20:2020: Economic Transformation Blueprint. Abuja: National Planning Commission.

70. Ogun, O. (1995). 'Country Studies: Nigeria', in S. M. Wangwe (ed.) Exporting Africa: Technology, Trade and Industrialisation in sub-Saharan Africa, UNU-INTECH Studies in New Technology and Development. London: Routledge, 246–95.

71. Oyelaran-Oyeyinka, B. (1997). 'Industrial Technology Policy: Making and Implementation in Nigeria: An Assessment'. NISER Occasional Paper. Ibadan: NISER.

72. Oyelaran-Oyeyinka, B. (2004). 'Networking Technical Change and Industrialization: The Case of Small and Medium Firms in Nigeria'. ATPS Special Paper Series No. 20. Nairobi: African Technology Policy Studies Network.

73. World Bank (WB) (2006). Nigeria Investment Climate Survey. Washington DC: World Bank.

74. World Bank (WB) (2014). Nigeria Country Overview, available at www.worldbank.org, accessed 25 June 2015.

75. Prof. ogbonnaya ofor,(2018). 'industrialization: key to national economic independence' dean, college of natural & applied sciences, copyright © 2018 veritas university Abuja.

76. Rowley, John. "Population", Microsoft® Encarta® 2007 [CD]. Microsoft Corporation, 2006.

77. ^ "Can Africa really learn from Korea?". Afrol News. 24 November 2008. Archived from the original on 16 December 2008. Retrieved 16 February 2009.

78. ^ "Korea role model for Latin America: Envoy". Korean Culture and Information Service. 1 March 2008. Archived from the original on 22 April 2009. Retrieved 16 February 2009.

79. ^ Leea, Jinyong; LaPlacab, Peter; Rassekh, Farhad (2 September 2008). "Korean economic growth and marketing practice progress: A role model for economic growth of developing countries". Industrial Marketing Management. Elsevier B.V. (subscription required). 37 (7): 753–757. doi:10.1016/j.indmarman.2008.09.002.

80. ^ Derek Gregory; Ron Johnston; Geraldine Pratt; Michael J. Watts; Sarah Whatmore, eds. (2009). "Asian Miracle/tigers". The Dictionary of Human Geography (5th ed.). Malden, MA: Blackwell. p. 38. ISBN 978-1-4051-3287-9.

81. ^ a b c "The 'paradoxes' of the successful state". European Economic Review. 41 (3-5): 411–442. 1997-04-01. doi:10.1016/S0014-2921(97)00012-3. ISSN 0014-2921.

82. ^ Chang, Ha-Joon. "The East Asian Development Experience".

83. ^ Data for "Real GDP at Constant National Prices" and "Population" from Economic Research at the Federal Reserve Bank of St. Louis.

84. ^ John Page (1994). Stanley Fischer; Julio J. Rotemberg, eds. "The East Asian Miracle: Four Lessons for Development Policy". NBER Macroeconomics Annual

1994. Cambridge, Massachusetts: MIT Press. 9: 219–269 [225]. doi:10.1086/654251. ISBN 9780262560801. Archived from the original on 2 February 2013.

85. ^ "Economic History of Hong Kong"Archived 17 April 2015 at the Wayback Machine., Schenk, Catherine. EH.net 16 March 2008.

86. ^ "Singapore Infomap – Coming of Age". Ministry of Information, Communications and the Arts. Archived from the original on 13 July 2006. Retrieved 17 July 2006.

87. ^ Thomas, Vladimir. "the World Transformed 1945 to the present". Michael H. Hunt. p. 352.

88. ^ a b John Page (1994). Stanley Fischer; Julio J. Rotemberg, eds. "The East Asian Miracle: Four Lessons for Development Policy". NBER Macroeconomics Annual 1994. Cambridge, Massachusetts: MIT Press. 9: 219–269 [247]. doi:10.1086/654251. ISBN 9780262560801. Archived from the original on 2 February 2013.

89. Woodford, Chris. (2008/2017) Technology timeline. Retrieved from https://www.explainthatstuff.com/timeline.html. [Accessed (August, 2018)]

90. Almond, Gabriel A. and James S. Coleman (eds.) The Politics of the Developing Areas. Princeton: Princeton University Press, 1960.

91. Bah, Abu Bakarr "Approaches To Nation Building In Post-Colonial Nigeria". Journal of Political and Military Sociology.

Http://Findarticles.Com/P/Articles/Mi_Qa3719/Is_2004 07/Ai_N9435086 2004.

92. Caroline Stephenson, "Nation Building", 2005, http://www.beyondintractability.org/essay/nation_building/

93. Iweriebor, Ehiedu, 1990 "Nigerian Nation Building Since Independence." Nigerian Journal of Policy and Strategy, Volume 5, Numbers 1 & 2. JACON.

94. Ibrahim A. Gambari, "Nigeria - The challenge of nation building and external relations" The Ado Bayero Lecture Series, Centre For Democratic Research ad Training, Bayero University, Kano, Nigeria, 8 February 2006

95. Pye, Lucian W. Aspects of Political Development. Boston: Little, Brown and Company, 1966

96. Suberu, R T. 1999 "Public Policy and National Unity in Nigeria". Ibadan: Development Policy Center

97. Wikipedia, "Nation building".

98. Thanks to my erudite sister, Abiola Saba (Timeless Impact) of Mantua, NJ who contributed in no small measure to this article.

99. ^ Karl Wolfgang Deutsch, William J. Folt, eds, Nation Building in Comparative Contexts, New York, Atherton, 1966.

100. ^ Mylonas, Harris (2017),"Nation-building," Oxford Bibliographies in International Relations. Ed. Patrick James. New York: Oxford University Press.

101. ^ Nairn, Tom; James, Paul (2005). Global Matrix: Nationalism, Globalism and State-Terrorism. London and New York: Pluto Press.; first used in James, Paul (1996). Nation Formation: Towards a Theory of Abstract Community. London: Sage Publications. See also James, Paul (2006). Globalism, Nationalism, Tribalism: Bringing Theory Back In —Volume 2 of Towards a Theory of Abstract Community. London: Sage Publications.

102. ^ Mylonas, Harris (2012). The Politics of Nation-Building: Making Co-Nationals, Refugees, and Minorities. New York: Cambridge University Press. p. 17. ISBN 1107661994.

103. ^ Keith Darden and Harris Mylonas. 2016. "Threats to Territorial Integrity, National Mass Schooling, and Linguistic Commonality," Comparative Political Studies, Vol. 49, No. 11: 1446-1479.

104. ^ Keith Darden and Anna Grzymala-Busse. 2006. "The Great Divide: Literacy, Nationalism, and the Communist Collapse." World Politics, Volume 59 (October): 83-115.

105. ^ Barry Posen. 1993. "Nationalism, the Mass Army and Military Power," International Security, 18(2): 80-124.

106. ^ Wimmer, Andreas (2018-07-04). "Nation Building: Why Some Countries Come Together While Others Fall Apart". Survival. 60 (4): 151 164. doi:10.1080/00396338.2018.1495442. ISSN 0039-6338.

107. ^ Mylonas, Harris (2012). The Politics of Nation-Building: Making Co-Nationals, Refugees, and

Minorities. New York: Cambridge University Press. ISBN 1107661994.

108. ^ Mylonas, Harris (2012). The Politics of Nation-Building: Making Co-Nationals, Refugees, and Minorities. Cambridge: Cambridge University Press. p. xx. ISBN 9781107020450. Retrieved 2013-12-02. Many journalists,

109. ^ Deutsch, Karl W. (2010). William J. Foltz, ed. Nation building in comparative contexts (New paperback print. ed.). New Brunswick [N.J.]: AldineTransaction. ISBN 9780202363561.

110. ^ Connor, Walker (18 July 2011). "Nation-Building or Nation-Destroying?". World Politics. 24 (03): 319–355. doi:10.2307/2009753.

111. ^ Jochen Hippler, ed. (2005). Nation-building: a key concept for peaceful conflict transformation?. translated by Barry Stone. London: Pluto. ISBN 0745323367.

112. ^ Smith, Anthony. 1986. "State-Making and Nation-Building" in John Hall (ed.), States in History. Oxford: Basil Blackwell, 228–263.

113. ^ Harris Mylonas. 2010. "Assimilation and its Alternatives: Caveats in the Study of Nation-Building Policies", In Rethinking Violence: States and Non-State Actors in Conflict, eds. Adria Lawrence and Erica Chenoweth. BCSIA Studies in International Security, MIT Press.

114. ^ Stephenson, Carolyn (January 2005). "Nation Building". Beyond Intractability. Retrieved 27 June 2018.

115. ^ Dobbins, James, Seth G. Jones, Keith Crane, and Beth Cole DeGrasse. 2007. The Beginner's Guide to Nation-Building. Santa Monica, Calif.: RAND Corporation.

116. ^ Darden, Keith; Mylonas, Harris (1 March 2012). "The Promethean Dilemma: Third-party State-building in Occupied Territories". Ethnopolitics. 11 (1): 85–93. doi:10.1080/17449057.2011.596127.

117. ^ Fukuyama, Francis. January/February 2004. "State of the Union: Nation-Building 101", Atlantic Monthly.

118. ^ Fukuyama, Francis (ed.) (2006). Nation-building: Beyond Afghanistan and Iraq ([Online-Ausg.] ed.). Baltimore, Md.: Johns Hopkins Univ. Press. ISBN 0801883342.

119. Engin, Kenan (2013). "Nation-Building" – Theoretische Betrachtung und Fallbeispiel: Irak(in German). Baden Baden: Nomos Verlag. ISBN 978-3-8487-0684-6.

120. Hodge, Nathan (2011), Armed Humanitarians: The Rise of the Nation Builders, New York: Bloomsbury USA.

121. James, Paul (1996). Nation Formation: Towards a Theory of Abstract Community. London: Sage Publications.

122. James, Paul (2006). Globalism, Nationalism, Tribalism: Bringing Theory Back In—Volume 2 of Towards a Theory of Abstract Community. London: Sage Publications.

123. Mylonas, Harris (2012). The Politics of Nation-Building: Making Co-Nationals, Refugees, and Minorities. New York: Cambridge University Press.

124. Mylonas, Harris (2017),"Nation-building," Oxford Bibliographies in International Relations. Ed. Patrick James. New York: Oxford University Press.

125. Smith, Anthony (1986), "State-Making and Nation-Building" in John Hall (ed.), States in History. Oxford: Basil Blackwell, 228–263.

126. Fritz V, Menocal AR, Understanding State-building from a Political Economy Perspective, ODI, London: 2007.

127. CIC/IPA, Concepts and Dilemmas of State-building in Fragile Situations, OECD-DAC, Paris: 2008.

128. Whaites, Alan, State in Development: Understanding State-building, DFID, London: 2008.

www.ingramcontent.com/pod-product-compliance
Lightning Source LLC
Chambersburg PA
CBHW021812170526
45157CB00007B/2551